Making Medicine

Surprising Stories from the History of Drug Discovery

KEITH VERONESE

Prometheus Books

Guilford, Connecticut

PB Prometheus Books

An imprint of Globe Pequot, the trade division of
The Rowman & Littlefield Publishing Group, Inc.
4501 Forbes Boulevard, Suite 200, Lanham, Maryland 20706
www.rowman.com

Distributed by NATIONAL BOOK NETWORK

British Library Cataloguing in Publication Information Available

Library of Congress Cataloging-in-Publication Data

Names: Veronese, Keith, author.
Title: Making medicine : surprising stories from the history of drug
    discovery / Keith Veronese.
Description: Lanham : Rowman & Littlefield, 2022. | Includes index. |
    Summary: "In Making Medicine: Surprising Stories from the History of
    Drug Discovery, author Keith Veronese examines fifteen different
    molecules and their unlikely discovery—or, in many cases, their second
    discovery—en route to becoming invaluable medications"—Provided by
    publisher.
Identifiers: LCCN 2021052258 (print) | LCCN 2021052259 (ebook) | ISBN
    9781633887534 (cloth) | ISBN 9781633887541 (epub)
Subjects: LCSH: Drug development. | Drugs—Research—History. |
    Drugs—Design.
Classification: LCC RM301.25 .V47 2022 (print) | LCC RM301.25 (ebook) |
    DDC 615/.19—dc23/eng/20211025
LC record available at https://lccn.loc.gov/2021052258
LC ebook record available at https://lccn.loc.gov/2021052259

∞™ The paper used in this publication meets the minimum requirements of American National Standard for Information Sciences—Permanence of Paper for Printed Library Materials, ANSI/NISO Z39.48-1992.

For Carla, Emelyn, and Ariana

Contents

Introduction

HOW DOES MODERN DRUG DISCOVERY WORK?

How do scientists design the pharmaceuticals we use to improve our lives? A handful are happy accidents and overlooked mixtures of carbon and hydrogen that become million- and billion-dollar makers for corporations. They also improve the lives of people the world over in the process. While the unintended discoveries in the chapters that follow may appear miraculous, drug discovery, under normal conditions, is a very laborious and rational undertaking. At the core of modern, rational drug discovery is the leveraging of multiple fields together to make one wondrous event happen—the binding of a molecule in the body to relieve a pain, erase an inconvenience, or cure a disease.

What Is the Target of a Drug?

Potential drug targets play a role in a disease state, but they do not have to cause the disease themselves. The positive modification of a drug target, in theory, should result in a biological change (e.g., the breakdown of a cell signaling pathway) that halts the disease or acute problem at hand. How does this modification occur? It's all about binding, the act of two structures, the drug (often a small molecule) and the target, coming together in hope that a promising result occurs. What is a small molecule? For the purpose of drug design, it is the product of a reproducible chemical reaction that scientists aim to be 500 daltons or less in size. A dalton is a unit of atomic mass that is one-twelfth the mass of a carbon atom. To give some examples, a molecule of water is a little more than eighteen daltons, sodium chloride (table salt) is fifty-eight and a half daltons, a molecule of the pain reliever ibuprofen is just over 206 daltons, and a molecule of the blood pressure drug losartan is just under 423 daltons. Ibuprofen binds to enormous cyclooxygenase enzymes, stopping the 70,000-dalton enzymes from doing their work and bringing pain relief in a matter of an hour. Not bad for a small molecule of 206 daltons.

Ligand-Based Drug Design versus Structure-Based Drug Design

Drug design often revolves around the testing of molecules in hopes of finding one that binds to and stops or alters the action of an enzyme or other biological target and results in a positive effect on the body. There are multiple approaches to drug design, with ligand-based and structure-based being at the core.

Ligand-based drug design lacks detailed knowledge of the structure of the biological target of a possible drug, relying only on previous known information about what binds the target. A ligand, by definition, is a molecule that binds to a larger molecule. In drug design, our ligand is a small-molecule drug candidate. By making rational modifications to other molecules known to bind the target, it is possible to create a more effective small-molecule drug even without knowing the exact structure of the target. This is difficult, however, as researchers are essentially flying blind as they synthesize a number of small molecules in hopes that one will perform superior to previous results.

Structure-based drug design is possible when highly detailed information about the shape and atomic composition of the drug target is known. This is often possible when high-resolution x-ray crystallography data exist, data showing the three-dimensional structure of the target through a mesh of electron clouds, or with high-quality nuclear magnetic resonance (NMR) spectroscopy. X-ray crystallography is an art to itself, relying on scientists to grow a crystal of the desired target at high concentrations of protein and then subjecting the fragile crystal to a barrage of x-rays, often at extremely low temperatures. It is not at all uncommon for a crystal to become damaged in the process, setting back months of work. This work is performed merely to obtain the raw data, which must then be meticulously processed before a three-dimensional structure of the crystalized target is revealed. Access to a well-resolved three-dimensional structure opens the doors to computer-aided techniques in drug design and also aids in the high-throughput screening techniques we will learn about next.

High-Throughput Screening and Its Role in Drug Design

High-throughput screening involves scanning libraries of thousands to millions of small molecules in an assay against an established target to see if they can alter the target in any way. An assay is just a fancy word for a test of any type that can qualitatively or quantitatively measure the change in activity of the target. Each small molecule in the library that successfully attains a minimum threshold for activity in the assay is called a "hit." These small molecules are often on the scale of 500 or less daltons. Successful hits are screened against other known targets as well to make sure they do not affect them and cause undesirable effects down the line. Due to the sheer size of the small-molecule libraries used in high-throughput screening, testing of compounds is often a job requiring significant automation, with unattended robotic systems capable of assaying and scanning the results of tens of thousands of compounds in a single day.[1] If multiple hits are similar in structure, it is likely that additional small molecules will be synthesized through organic chemistry methods with well-known bioisosteric alterations in hope of further increasing the potency against the target. A bioisostere is a group of atoms that behaves similarly either physically or chemically in biological organisms to another set of atoms. This is part of leveraging the structure–activity relationship between the small molecule and the target to create a better binding event.

High-throughput methods can also take place in silico, where small molecules in a digital library are spun and manipulated to see how well they will "dock" with a computer-generated model of a known target. These computer-generated models take into account the amount of space present for the small molecule to attach using chemistry concepts like electron density and hydrogen bonding. The computer-generated models are often informed by earlier successes in x-ray crystallography and NMR spectroscopy. Such tests are extremely processor intensive and are imperfect, but still yield quality results as to what type of molecules may bind to the drug target before any benchtop chemistry or assays are performed.

Akin to high-throughput screen is fragment-based lead discovery, which relies on smaller libraries of compounds (usually in the thousands) and even smaller molecules than used in a high-throughput screen (often less than 300 daltons). These smaller molecule fragments are often successful in binding hard to reach "hot spots" on the target that are impossible for the larger (but still small) molecules used in high-throughput screening to target.[2] Once a substantial number of binding fragments are discovered through a series of assays, the fragment can be "grown" to the size of a typical small-molecule drug through rational organic synthesis and the addition of atoms (or functional groups if you have sat through any organic chemistry lectures) that will further aid in the reception of the drug candidate. Also available is the possibility of combining well-performing fragments, especially if their binding to the target overlapped in multiple spaces, in a process scientists versed in fragment-based lead discovery call "merging." This is one key advantage of fragment-based lead discovery over high-throughput screening, as the two "merged" fragments will still be on the order of less than 500 daltons, approximately the size of a small-molecule drug candidate. There could also be multiple ways to merge two fragments, allowing for multiple small molecules to come out of such a fragment-based study.[3]

One may start with a library of a million small molecules to arrive at a single lead compound that will be rigorously tested through additional in vitro studies and, as the goal, animal and human clinical trials in the process of drug development. The enormity of this task makes the unintended discovery of the drugs to be discussed later in the book all the more fascinating and serendipitous.

The Role of Natural Sources in Drug Design

Nature is a bountiful source of possible pharmaceuticals due to its diverse environments and the sheer number of defense mechanisms organisms develop to survive. Medicinal chemists, armed with knowledge of the role that specific molecules play in a plant or microbe or even a rare invertebrate that inhabits the bottom of the ocean, often use these identified molecules as starting points for solving similar problems in the human body. For example, a crude extract of willow tree bark has been used for thousands of years to alleviate pain and fever. Throughout time, scientists discovered that the active ingredient in the willow bark extract was a small molecule now known as salicylic acid, which played a role in the immune response mechanisms of the tree. By slightly altering salicylic acid, scientists created acetylsalicylic acid, better known to us as aspirin. Thankfully, we have aspirin to take when a headache or fever arrives and are not left to scour the woods for a willow tree and subsequently concoct a no doubt bitter elixir.

A more recent (and exotic) example is the drug ziconotide (brand name Prialt), which is derived from a toxin present in *Conus magus*, the cone snail, a venomous sea-dwelling creature that can kill with a sting. The result is a pain reliever quite strong—1,000 times stronger than morphine—yet so delicate that it must be delivered directly into the spinal fluid so as to not be blocked by the blood–brain barrier.[4]

What Is Bioavailability?

Bioavailability is key to drug design. It is the measure of the amount of drug actually in the bloodstream and open to use by the body compared to the total dose taken. Why would these two be different? Most drugs are taken orally and thus subject to the gastrointestinal tract. This is not all bad, as the pills do need to be dissolved before active ingredients within are released. But within the gastrointestinal tract, there are a number of options for the drug to be metabolized or modified before it reaches the bloodstream and the drug's desired place of action. Also, if the drug passes through the liver, there is a significant chance for the available, bioactive fraction of the drug to be reduced prior to it reaching systemic circulation. This happens as the liver further metabolizes the drug or excretes it through bile in what is known as the first pass effect.[5]

One way to get around a significant decrease in bioavailability is to introduce the drug intravenously, but this isn't exactly an option when you go to pick up your prescriptions at your local pharmacy. Ensuring the ability of a drug to be taken orally is a key thought in the back of the minds of chemists and pharmacologists throughout the design phase, as it increases the ease of taking the medication and does not introduce need for administration by injection, which may limit the market for the drug.

What Is Lipinski's Rule of Five?

Lipinski's Rule of Five is a quick method for medicinal chemists to get an idea of a drug candidate's effectiveness and whether it will be orally active. It is often used at the earliest stages of drug development, typically when target binding studies are still taking place. Interestingly, there are only four "rules," as "five" relates to the fact that all of the rules have a multiple of five within them. Lipinski's Rule of Five states that a drug will have a better chance at being orally active if it has one or fewer of the following: (1) five total hydrogen bond donors (a measure of whether a molecule will loan a hydrogen to another), (2) ten total hydrogen bond acceptors (a measure of whether a molecule will be attracted to a hydrogen on another molecule), (3) a molecular mass of greater than 500 daltons (emphasizing the small-molecule nature of most drugs), and the admittedly difficult to understand (4) octanol–water partition coefficient greater than five. The latter is a measure of the small molecule's ability to dissolve into lipids (i.e., fats) and become entangled in various organ systems as it journeys through the body.[6]

What Is a Biologic Drug, and How Are They Different from Small-Molecule Drugs?

So far, our discussion has revolved around small-molecule drug design. These small molecules are reliably reproduced through well-known organic synthesis reaction techniques. But there is another very different form of pharmaceutical that makes up a large amount of the medications on the market—the biological product, or biologic for short. Biologics meet the patient in a variety of forms, whether it be as a vaccine, protein formulation, stem cell treatment, monoclonal antibodies, or even blood components. As such, they are much more delicate, for lack of a better term, than synthesized small-molecule drugs. For example, taking a biologic orally would immediately render it useless, as biologics are made from substances your body is designed to break apart and digest, like proteins. Because of this, biologics are made available through a variety of other means, including injections and infusions. Common biologics on the market include the monoclonal antibody Humira and the fusion protein Enbrel, both of which treat rheumatoid arthritis. Nasal administration of biologics is also possible and aggressively researched, with a live-attenuated form of the influenza vaccine already available for intranasal delivery.[7]

Biologics are also much different structurally than your average small-molecule drug, often dwarfing them in size. As they are not taken orally, biologics do not need to follow tried-and-true methods of drug design like Lipinski's Rule of Five, nor are they typically exposed to the gastrointestinal tract or liver before reaching circulation in a struggle for bioavailability.

In lieu of the steady stream of organic synthesis reactions used to create small-molecule pharmaceuticals, companies pursuing biologics make use of living cells in the synthesis process by inserting a specific string of DNA coding for the desired biologic into a living cell. These cells can be from bacteria like *Escherichia coli*, for example, or from mammalian, yeast, plant, or viral cells. Creating the previously mentioned specific DNA is a master task in and of itself, with the newly created DNA coding for the biologic of choice called recombinant DNA. Once the DNA is inserted, the cells go about their business, replicating and producing the biologic in large quantities after it integrates the recombinant piece of DNA. Once enough cellular growth occurs, scientists "harvest" the cells, removing the biologic from the cells through an often-complicated process of purification.

Biologics are commonly temperature and condition sensitive (e.g., many are light sensitive), so much care is taken in their storage from this point on. The complicated nature of creating, isolating, and storing a biologic also leads to their much higher cost when compared to an average small-molecule drug.

Another large difference between small-molecule drugs and biologics is seen in examining the fidelity of their replication. Since biologics are the products of living cells, they can exhibit small variations from batch to batch. Subsequent batches of small-molecule drugs, on the other hand, are exact within quality assurance guidelines. The lack of a means to exactly reproduce a biologic plays a role in their approval, in their regulation, and when it comes time to produce a generic form of a biologic, which are called biosimilars.

What Was the First Synthesized Pharmaceutical?

Although the first pharmacy is believed to have opened in Baghdad in 754 CE, the malady-relieving medications sold at this time consisted of extracts of plant and animal matter.[8] Our modern approach to pharmaceutical synthesis would not come for another 1,000-plus years until the birth of organic chemistry, and even then, the therapeutic application had to wait a couple of decades. Justus von Liebig, a German scientist who is often viewed as one of the fathers of organic chemistry, published a paper on the synthesis of chloral hydrate in 1832.[9] To Liebig, this paper existed solely as an educational tool.

Three decades later in 1862, Johann Liebreich, a German physician, noted the sleep-inducing effects of chloral hydrate, along with patients noting the disgusting smell and taste of the syrup. Still, the wonderful sleep it brought about without significant drowsiness after waking up kept chloral hydrate a popular sedative and treatment for insomnia through World War II. Chloral hydrate was deemed safe for use in children as well, at least at that time.[10] The safety of chloral hydrate, however, is questionable from a modern perspective, as a combination of alcohol, the sedative pentobarbital, and chloral hydrate led to the premature death of Marilyn Monroe.[11] The addition of chloral hydrate to alcohol is particularly dangerous, with the laced drink going by the slang name "Mickey Finn." Slipping someone a "Mickey" was popularized in Chicago's Whiskey Row during the 1890s thanks to the drink's ability to knock a thief's mark unconscious.[12]

Although we could heap coals on the head of Liebig for missing out on this application of his own research, Liebig is an interesting character in his own right. His interests unfurled into a variety of fields from synthetic chemistry to the application of chemistry concepts in agriculture to show the value of nitrogen in fertilizer to the chemistry of food and particularly meats. Liebig founded a company to sell meat extracts and promoted "meat teas" for their perceived nutritional value.[13] Liebig also found time to rigorously investigate spontaneous human combustion, which was then viewed as the plight of drunkards. After viewing fifty reported cases of spontaneous human combustion, Liebig soaked cadaver tissue in 70 percent alcohol solutions and set fire to them to show that the alcohol burns off but that the tissue does not burn. Liebig also carried out some rather inhumane experiments on rats to further prove the point, injecting the rats with ethanol and then attempting to set them on fire.[14]

Legislating the Sale of Medicine in the United States

Legislation of the pharmaceutical industry moved relatively slowly in the United States. President Theodore Roosevelt signed the 1906 Pure Food and Drug Act, which banned the interstate and foreign trafficking of mislabeled food and drug products. The act also imbued the U.S. Bureau of Chemistry with the power to monitor and seize these products.[15] The 1906 Pure Food and Drug Act outlined eleven dangerous ingredients, including heroin and cocaine, which must be indicated and with the correct amount noted on the label if included in a food or drug.[16] The U.S. Bureau of Chemistry became the United States Food, Drug, and Insecticide Administration in 1927 en route to becoming what we know as the Food and Drug Administration (FDA) in 1930.[17] The year 1938 and the Roosevelt administration brought another key piece of legislation, the Federal Food, Drug, and Cosmetic Act, which gave the FDA oversight of just that—food, drugs, and cosmetics—and, later, medical devices.[18] The 1938 act also called for all drugs sold to come with labels for proper use and decreed that all new drugs must be shown to be safe before the FDA prior to being sold.[19]

The divide between prescription and over-the-counter pharmaceuticals was created in the United States in 1951 with the Durham–Humphrey Amendment, which defined anything addictive or possibly harmful and thus unsafe for self-medication to require a prescription to obtain.[20] This 1951 amendment was followed a decade later by the Kefauver–Harris Amendment of 1962, which required drug manufacturers to provide proof of the safety and effectiveness of a drug before obtaining approval for sale in the United States. The Kefauver–Harris Amendment originated out of the thalidomide birth defect tragedies of the late 1950s and early 1960s, tragedies stemming from women taking the popular European sedative to ease the symptoms of morning sickness. Thankfully, thalidomide did not receive approval for use in the United States, with an FDA reviewer, Frances Oldham Kelsey, refusing to approve the drug when it came across her desk in 1960. Kelsey bore concerns that thalidomide could cross the placental barrier due to research she did during World War II on antimalarial drugs that did just that.[21]

These four acts and amendments, in particular 1938's Federal Food, Drug, and Cosmetic Act, 1951's Durham–Humphrey Amendment, and 1962's

Kefauver–Harris Amendment, formed the kernel that is now the FDA process for approving new drugs—proof of safety and efficacy.

How Does a Pharmaceutical Obtain Approval by the FDA?

So how does a pharmaceutical company bring a new drug to market? This process takes years and often incurs billions of dollars of costs. The goal of this undertaking is a new drug application (NDA), but this only comes at the end of several rounds of testing to show that the drug will be safe and effective for use in humans. It is important to note here that the drug companies never have to know the mechanism of action of how a drug works to receive approval; they just have to show that the drug is safe and effective.

Prior to the filing of an NDA, several sets of studies are performed to gauge the effectiveness and safety of a drug candidate. The first of these studies takes place at the in vitro level, and if this is successful at the cellular level, an investigational new drug application (IND) is filed before moving on to the next step. The IND contains supporting research showing the drug candidate to be safe when tested in animal studies along with how the company manufactures the drug. The purpose of the IND is to rule out issues of extreme toxicity before the company moves on to clinical trials and also to show the FDA how the company plans to move through such trials, which is a daunting task and often leads to the death of many proposed drugs. Clinical trials make use of humans as the test subjects, starting with a small number in phase 1 and increasing in number through to post-marketing trials in phase 4.[22]

Phase 1 is small study, made up of twenty to 100 healthy volunteers or, for cancer drugs, twenty to 100 individuals who currently suffer from the cancer the drug aims to treat. This study phase often lasts several months, with the goal of determining the safety of the drug and the correct dosage in humans. The drug's effectiveness in treating a disease is not a major concern in this phase. Side effects are closely monitored with increases and decreases in dosage, all part of an FDA mandate to observe how the drug candidate responds once the drug is in the human body. If the side effects are deemed too significant, clinical trials stop here. The FDA estimates that 70 percent of drug candidates "pass" phase 1.[23]

Phase 2 expands to a clinical trial of several hundred people who are currently suffering from the disease state the drug aims to treat. This phase can last months to up to two years, and once again, the key aim is monitoring the safe use of the drug in prospective patients. Effectiveness, however, can also be quantified to a small degree—due to the relatively low number of people in the trial—as you monitor the progress of your clinical trial population and their disease state. Phase 2 is often brutal, as only an estimated 33 percent of drug candidates move on from here.[24]

We should take a moment before we move to phase 3 to mention the sheer enormity of people necessary to conduct these trials apart from the clinical trial participants themselves. Chemists and microbiologists are necessary to create and manufacture significant quantities of the drug candidate; statisticians, pharmacists, and pharmacologists are present to interpret the drug's performance; nurses and physicians are necessary for administration of the drug; and medical officers are there to see that the whole process stays on the rails. As the size of these trials expands, it is easy to see how bringing a new drug candidate to market is such an expensive task let alone envision the sunk costs incurred for those drugs that fail in any phase of clinical trials.

Phase 3 studies expand from phase 2 and range from 300 to 3,000 individuals with the targeted disease state in double-blind, placebo-controlled trials and often last from anywhere from one to four years. This phase determines whether the drug candidate is effective and truly works for its intended purpose. The increased time span of phase 3 trials allows for long-term and rarer side effects to rear their heads and allows for a determination as to whether the drug candidate is safe and effective enough to take to market. The FDA estimates that between 25 and 30 percent of drugs successfully complete phase 3.[25]

If a pharmaceutical company is successful in moving through phase 3 clinical trials, the company can now file the aforementioned NDA. If you notice, several years have passed since the early preclinical and animal trials as the drug travels through the phases of clinical trials, putting on full display the rigorous nature of the FDA's approval process.

The FDA describes the NDA as the "full story of a drug." Within this story are data earned through the clinical trials about safety and effectiveness as well as information as to the drug candidate's possible abuse along with directions for proper use and proposed labeling. The FDA then assigns a team of statisticians, medical officers, inspectors, pharmacologists, and project managers to inspect the NDA throughout the course of six to ten months and vote thumbs up or thumbs down on the NDA approval. If a dreaded thumbs down is received, a pharmaceutical company does retain the right to appeal or to conduct further tests and trials until the FDA is satisfied.[26] Once the NDA is approved, the pharmaceutical company is free to manufacture the drug candidate for sale.

It is important to note that biologic pharmaceutical candidates follow a similar pathway through clinical trials and approval, with a biologics license application (BLA) being made in addition to an NDA. The approval of biologics is handled by the Center for Biologics Evaluation and Research, a subsection of the FDA.[27]

A phase 4 clinical trial does exist, but it takes place only after the drug is successfully released on the market. The phase 4 trial takes advantage of the subject base taking the drug in the population at large to thoroughly monitor the drug's safety and effectiveness.[28]

The Role of Academia in Drug Design and Development

The overwhelming majority of drug design research and development takes place in the private sector; however, partnerships between academic research institutions and pharmaceutical companies are not unheard of. These relationships can be tricky, as they often involve closely held patent information that bars or slows the publication of academic manuscripts coveted by professors and PhD students alike, with these publications representing the currency of their training and career advancement. Academic drug discovery is commonly funded in part in the United States through taxpayer-funded grants by the National Institutes of Health (NIH), in particular the NIH Roadmap Initiative, which seeks to fund high-risk yet high-reward research in the allied biomedical fields. In this research, university laboratories are often seeking out lead molecules to become drug candidates, which are then leveraged by pharmaceutical companies for further testing and, it is hoped, clinical trials. For university professors interested in tackling drug discovery, this is a wonderful partnership, as a university laboratory lacks the infrastructure or manpower to carry out the clinical trials needed to bring a drug candidate to market. When an academic collaboration provides a successful drug candidate, the academic institutions and laboratories are often rewarded by the pharmaceutical companies with lucrative licensing deals.[29] Although rare, several academic institutions have shown repeated success in drug discovery through partnership with pharmaceutical companies, leading to whole departments centered on the venture, including Emory University's Institute for Drug Discovery, St. Jude's Children's Research Hospital's Chemical Biology and Therapeutics Department, and the University of Florida's Center for Drug Discovery.[30]

How Do Pharmaceutical Companies Scale Up Production Once Approved?

The chemical engineering behind scaling up for production of a newly approved pharmaceutical could fill a book in and of itself. To give a short overview, pharmaceutical companies carry out manufacturing "campaigns" wherein they proceed with the mass production of a single drug at a time before cleaning up and moving on to produce the next one. During these campaigns, careful decisions are made to maximize the efficiency of the process to maximize profit while also reliably producing a quality product. For small-molecule drugs, the reactions between solids and liquids take place in 1,000-liter vats that are stirred vigorously and are temperature controlled. The temperatures are vitally important, as a small change in temperature can lead to the death of a reaction or the synthesis of an incorrect product, wasting substantial amounts of time and money. Multiple reactions likely need to be strung together to create the final desired drug, with filtration steps throughout to ensure purity. In the case of a drug that will be delivered as a pill or a tablet, the final active drug product is usually a powder, with additives (called excipients in the pharmaceutical industry) introduced to act as binders, fillers, flavoring, antioxidants, or preservatives. Before the final pill or tablet is created, the mixture of active and inactive ingredients is often milled to decrease the average particle size and aid in ensuring a homogeneous mixture that can be easily acted on by the body. Additional steps are needed if the final drug form is a liquid. For example, any pharmaceuticals taken via the eye have to be extensively sterilized to prevent the introduction of bacteria into such a sensitive area. Throughout a manufacturing process, numerous safety hazards are presented for those working in the synthesis campaign, from exposure to solvent fumes, both hot and cold reactor vats, and numerous "pinch points" in machinery.[31] Long-term exposure to the drug itself in the manufacturing process can also pose a health concern for the workers, as shown in an investigation on the effects of the manufacture of estrogens on male and female workers. In this case, the main source of exposure came through the inhalation of powdered, pure estrogen at a number of steps in the production process.[32]

The Drug Patent System in the United States

In the United States, a pharmaceutical company typically controls a drug for twenty years. This sounds like a long time, but in the eyes of the pharmaceutical world, this twenty-year span becomes a race against the clock to recoup as many dollars as possible out of a drug before patent expiration. Drugs are often patented long before the first clinical trial takes place. Once the new composition-of-matter patent is filed with the U.S. Patent and Trademark Office, the countdown is on, with every phase of the clinical trial process counting further and further down the clock. Once the NDA is approved by the FDA, the drug is also then listed in the FDA's *Approved Drug Products with Therapeutic Equivalence Evaluations*, known by the slang term *Orange Book* in pharmaceutical circles, along with any patents claiming to support this approval. Patent protection varies from country to country, but for the scope of this book, we will stick with the U.S. process due to the sheer number of drugs developed in the country.

Often, by the time drug development and the FDA's new drug approval process are concluded, the pharmaceutical manufacturer may hold only seven to ten years of patent protection left with which to recoup the billion-dollar-plus expense of bringing a new pharmaceutical from cradle to market.[33] During this time, no other company can sell the same drug on the market in the United States, and if they did so, the company would be opening themselves up to a world of litigation.

Patent protection is not a bad thing, however, as it is at the core of why so many drug companies are interested in conceiving of new pharmaceuticals. If patent protection did not exist, there would be zero to low incentive for a pharmaceutical company to create and formulate new drugs. If patents did not exist, all of the time, money, and effort spent in the research and development phase as well as the FDA approval process could be rendered moot the moment a company's new drug came to market and a number of copycat competitors with their own version arrived on the scene months later. It is of note that only 14 percent of drug candidates—although this varies greatly across drug types—that reach the clinical trial phase gain approval by the FDA.[34] Clinical trials, including the unsuccessful ones, can easily cost $100 million or more. Add to this the additional millions of dollars necessary for research and development on a drug, and the bill quickly becomes a billion-dollar price tag, all without a guarantee of FDA approval.[35] Patents on drugs protect a pharmaceutical company's own investment while indirectly ensuring that the patient base receives access to a steady flow of new therapies coming onto the market. Pharmaceutical companies are not manned by saints, however, as it is common for the price of a brand-name drug to go up in the six to eighteen months prior to patent expiration as a last-ditch effort to earn as much as possible from the brand-name version.[36]

Patent Extension through Altered Form or Altered Delivery Method

Although there are means by which to squeeze the most out of the twenty-year patent clock regardless of time spent in the FDA approval process, drug companies are wise enough to seek out additional methods to start the patent clock over again. To eke out additional money from a drug that a pharmaceutical company is about to lose patent control over, it is common for an altered form of the drug to be released, with the new form carrying its own patent protection. The creation of an extended-release version of an existing drug is a popular way to regain some of the market and still provide some benefit to the consumer through a steady release of medication through the day. Other forms include patenting a drug formulation containing the active metabolite—a metabolite is a molecule modified by the processes of the body—of an already existing drug. For example, Wyeth Pharmaceuticals' brand name Pristiq (generic name desvenlafaxine) is a small-molecule drug made up of the active metabolite of parent company Pfizer's Effexor (generic name venlafaxine), with both molecules functioning as serotonin-norepinephrine reuptake inhibitors.[37] Another example comes from separating out isomers in an existing mixture found in a drug currently success-ful and on the market. The antidepressant Celexa (generic name citalopram) is made up of left-handed and right-handed isomer forms of citalopram, with your body metabolizing and making use of only the left-handed isomer for the anti-depressant effect.[38] As the patent on Celexa neared expiration, Lexapro (generic name escitalopram) was introduced in 2002, with Lexapro containing only the active left-handed isomer.[39] AstraZeneca did the same with its popular gastro-esophageal reflux disease medication Prilosec (generic name omeprazole), which is a racemic mixture of two stereoisomers. In this case, however, both isomers are useful, as the right-handed and the left-handed forms are converted into the same molecule by the body thanks to the acidic conditions of the parietal cells in the gastrointestinal tract.[40] The pharmaceutical giant later marketed just the left-handed form as Nexium (generic name esomeprazole). Such small changes are major moneymakers for pharmaceutical manufacturers, as they often require very little additional research and development.

A pharmaceutical company may try yet another method and bypass the prescription process altogether. Pharmaceutical manufacturer Schering-Plough showed its popular brand name prescription allergy medicine Claritin to be safe enough for the average individual to dose at prescription strength, a telltale sign that a drug might be able to be sold without a prescription. This allowed Schering-Plough to move forward in 2002 with approval for an over-the-counter version of Claritin as the patent period neared expiration.[41] A slight change in the way Claritin was sold allowed the brand to remain strong while reducing the price and giving the consumer an unobstructed pathway to the product. It did not hurt that it provided Schering-Plough with an additional revenue stream as one was about to dry up.

What Portion of the World's Pharmaceutical Research Is Carried Out in the United States?

Using patents acquired as the key metric, a retrospective study from 1996 to 2013 of 556,112 patents granted worldwide showed U.S. entities to acquire

126,747 patents pertaining to pharmaceuticals. This fares well compared to the total patent output of Europe at 162,721, with Europe led by the contributions of Germany (32,534 patents) and France (23,904 patents). Not surprisingly, Japan and China also fared well, with China acquiring 39,460 patents to Japan's 33,539 in the same time period. Of particular interest is the rise of China in the patent-acquiring game, with the country going from forty-one patents acquired in 1996 to 1,218 patents in 2000 and then again increasing to 2,165 patents in 2005 and 4,352 in 2013. This increase in Chinese intellectual property acquisition corresponds with two separate five-year plans of government support for the pharmaceutical industry at the tail end of the twentieth and the beginning of the twenty-first century. If current trends in patent acquisition continue, China stands to surpass the annual patent output of the United States and Europe by 2036.[42]

So, the raw data suggest that the United States and Europe, when combined, carry out the lion's share of the patent-acquiring research in the world. This posits the United States as a significant leader, with the United States tripling the output of the next largest patent-producing country, China, during this seventeen-year time span. But will this continue, or is the United States merely the benefactor of a head start? We will have to wait and see to find out.

A long-standing dogma persists that the United States pays more per person for pharmaceuticals due to the free market nature of its economy and therefore that companies within its borders are enabled to innovate and create more new pharmaceuticals. Whether this is true is difficult to untangle, but it does appear that innovation thrives in the United States, at least for now.

Where Do Generics Enter the Game?

What happens when the patent on a drug expires? This is a big event in the life of a medication. The U.S. patent system works in a manner so as to give the patent holder a right to exclude others from infringing on their patent in exchange for the patented material entering into the public domain once the patent term expires. As such, once a patent expires, other pharmaceutical manufacturing companies are then allowed to make use of the previously protected intellectual property and manufacture so-called generic forms of the brand-name small-molecule drug. The pathway for a generic drug to enter the market is outlined by the Hatch–Waxman Act of 1984, also known as the Drug Price Competition and Patent Term Restoration Act of 1984. Orrin Hatch of the state of Utah is the "Hatch" of the Hatch–Waxman Act, with the longest-serving Republican senator in U.S. Senate history still playing a key role in shaping the generic market nearly four decades later.

The Hatch–Waxman Act of 1984 held two key provisions aiding the arrival of generic medications onto the market. The first allowed for a simpler, abbreviated new drug application (ANDA) to be filed. This new form of drug application required the generic manufacturer only to show how it will proceed in synthesizing and scaling up the manufacturing of the drug, along with a series of quality assurance tests and experimental data showing the generic form to act the same in the human as the brand-name drug. This was a boon for the generic market and required far less than needed—particularly no clinical trials—to formulate the NDA required to bring a brand-name drug to market. The generic manufacturer is allowed to perform the tests needed to acquire an ANDA prior to the expiration of the patent on the drug of choice due to a statutory exemption against patent infringement in the Hatch–Waxman Act. This greatly decreases the time it takes for a generic to reach the market after a brand name's patent expires.[43] The second key provision of the Hatch–Waxman Act of 1984 gave an exclusivity period to the first generic manufacturer to challenge an expiring patent listed in the FDA's *Orange Book* with a generic drug, giving the generic manufacturer a significant head start on market share and price setting for the generic forms of the newly off-patent drug.[44] This short period of exclusivity does not last forever, ending at 180 days. When the 180 days are up, the price of a generic drug begins settling and eventually falling considerably throughout the years as more and more generic manufacturers begin producing their own versions, with market competition winning out for the consumer's benefit. A 2019 FDA study showed that with a single generic manufacturer, prices drop roughly 40 percent, but with two generic manufacturers, they are slashed to less than half of the brand-name price. When the number of additional manufacturers increases past six, we see a strong force multiplier, as the cost of the generic often approaches 5 percent or less than the brand-name version.[45]

Generics do not exist for biologic products due to the variability from batch to batch of a given product during the manufacturing process and the inability to make an identical copy of the reference biologic. Instead, the off-patent form of a biological drug is deemed a "biosimilar," a tag that carries with it a necessity to perform the same as the brand-name "innovator" biologic. Biosimilars are also required to undergo further clinical testing not required for a generic small-molecule ANDA before they are approved.[46] This probably goes without saying, but these additional constraints make it much more difficult and cost intensive for a generic manufacturer to produce a biosimilar than a given generic small-molecule drug.

While the United States and Europe are hotbeds for drug research and development, generic manufacturing thrives across the globe. Industry heavyweight Teva Pharmaceuticals hangs its hat in both Petah Tikva, Israel, and Parsippany, New Jersey. Mylan is located in the Netherlands, the United Kingdom, and Pennsylvania, while Sandoz is a part of the Swiss pharmaceutical mega-power Novartis. Apotex, a company that launched one of the first biosimilars, is helmed in Toronto, Canada, with facilities in Winnipeg as well. Dr. Reddy's Laboratories, Sun Pharma, and Aurobindo Pharma all call India home. Generic manufacturing opens up access to pharmaceuticals to billions across the globe due to lower prices while also providing a steady supply of jobs across the world.

The Hatch–Waxman Act carried with it an important third provision, this one aiding manufacturers of brand-name pharmaceuticals instead of generic manufacturers. This provision allowed for an up to five-year extension in patent rights to cover a portion of the time a hoped-to-be brand-name drug is under review by the FDA. To gain an extension of this sort requires the brand-name manufacturer to list all accompanying patents associated with a newly approved drug in the FDA's *Orange Book*.[47] Most Hatch–Waxman Act–related patent extensions are on the order of three years, but any precious time recouped leads to additional cash flow for the innovator of the pharmaceutical.[48]

The generic market is not without confusion, however. A cornerstone of confusion in this arena is when brand-name manufacturers seek to enter the generic market with the creation of their own "authorized generics." Authorized generics often cut into the profits of the first generic manufacturer to reach the market with their own version and allow the brand-name company to increase their revenues (albeit not at the same levels as when the innovating manufacturer held exclusivity) as individuals switch to the generic to take advantage of their often-lower cost.[49] Another trick up the sleeve of pharmaceutical manufacturers is reverse payment agreements, wherein a pharmaceutical company pays a generic manufacturer not to produce the generic or to delay production. AstraZeneca took part in one of the more publicized reverse payment agreements in recent history when it paid Ranbaxy Laboratories $700 million to delay the production of a generic form of AstraZeneca's popular acid reflux medication Nexium.[50] Such astronomical numbers give us an idea of just how much money is at stake for both the innovating manufacturer clinging to the rights of the brand-name drug and a generic manufacturer waiting to take up the mantle.

It is now hoped that you are armed with some new knowledge about how modern, rational drug discovery takes place along with a number of the hurdles pharmaceutical companies have to overcome to bring a medication to market. With this in hand, let's take a look at how a number of vital pharmaceuticals have been discovered in the past, beginning with the antibiotic penicillin.

CHAPTER 1

Penicillin

The structure of Penicillin F. One of the many versions of penicillin, with all reliant on their peculiar ring structure for their activity. *Structure generated by author in ChemDoodle v 11.5.0, using information supplied by the National Institutes of Health*

Penicillin is one of the wonders of the twentieth century and our faithful friend in the struggle against the invisible threats of bacteria and infection, a friend that paved the way for an army of antibiotics that saved the lives of hundreds of millions in the process. The story behind penicillin is just as interesting as the drug is lifesaving. It is the story of a young man frustrated by the medical treatment of his fellow soldiers, the story of a remarkably keen eye for observation, the story of a man unable to make his vision a reality, the story of a cadre of researchers driven to find a cure for bacterial infections amidst the backdrop of World War II, and a story laden with so many crucial points where we could have lost any hope of making any use of the drug forever. Let's start our story with a young

1

Alexander Fleming and his circuitous route to the world of laboratories, mold, and bacteria.

Alexander Fleming's Inquisitive Mind

Born to farmers, Alexander Fleming did not begin graduate medical training right after finishing his formative studies. Instead, he carried out the mundane occupation of working in a shipping office for four years (I wonder if any of his eye for detail or exceptions arose during this time period or if it was always there, waiting for opportunities to show). A myth persists that Alexander Fleming (or his father, depending on the variant), while toiling on the farm, saved Winston Churchill from drowning in a bog. The tale goes on to say that Churchill's father, the Lord Randolph Churchill, sponsored the younger Fleming's medical school education as a thank-you, but this is a whimsical construction from a 1950s children's church program.[1] Instead, Fleming's deceased uncle John willed a share of his estate to him, and in becoming his benefactor, the late John Fleming enabled Alexander Fleming to attend St. Mary's Medical School at London University and forever changed the world.[2]

Graduating at the age of twenty-five, Fleming drew high praise for a wholly different pursuit during this time—his marksmanship. Not wanting to lose one of his prize team members to graduation and the pursuit of the medical field, the rifle team's captain introduced him to Sir Almroth Wright, a researcher at St. Mary's Medical School, on the supposition that Fleming would stay on the rifle team if he was active in a position at St. Mary's.[3] The rifle team captain supposed correctly, with Fleming and Sir Almroth Wright working together for a large portion of their careers.

St. Mary's added Alexander Fleming as a lecturer in 1914 in part due to his work with Wright, but Fleming found himself born at an interesting time, a time that saw him called to duty as a member of the Royal Army Medical Corps throughout World War I.[4] One might think these four years cost Fleming invaluable research time, but Fleming made the most of the moments in between the chaos and gore, noting the sheer number of his countrymen battling infected wounds instead of the Central Powers. Fleming paid particular attention to the use of antiseptics against infection during the war, and he was not pleased.

Soldiers used flavine as an antiseptic in World War I, and Fleming no doubt saw the aftereffects of its use firsthand.[5] In his 1917 paper "The Physiological and Antiseptic Action of Flavine," Fleming fought a war against its use as an antiseptic, on the battlefield or otherwise. Antiseptics stop or inhibit the growth of bacteria, whereas antibiotics kill bacteria. Fleming, noted as a lieutenant of the

Royal Army Medical Corps in the pivotal paper published in the journal *The Lancet*, rails against the use of flavine as an antiseptic for open wounds. Flavine was thought to be the near ideal antiseptic during the era, particularly for wartime applications. Fleming, through a series of clever experiments, showed flavine to react violently with leukocytes (white blood cells), a key part of the human body's immune system, noting that the flavine killed leukocytes nearly as well as it stopped bacteria. Fleming also put an end to the idea of flavine being a useful intravenous antiseptic for blood infections before moving onto the paper's heart, wherein he rips apart the use of flavine as a topical wound antiseptic. Fleming does so meticulously and thoroughly, hypothesizing that flavine's action on a cleanly dressed wound would be detrimental to the soldier. Fleming showed flavine to kill off a portion of the body's immune response through the destruction of leukocytes while at the same time flavine failed to penetrate lower into the body to the bacteria under the wound, which are then left to flourish. In the end, Fleming deems the current use of flavine as a panacea for septic wounds "unsound."[6]

Early Success with an Antibacterial Agent

After the end of World War I, Fleming belonged to St. Mary's once again, where he would continue his research with a keen eye pointed to tackling human-harming infectious bacteria. In 1922, Fleming publishes what most would consider the work of a lifetime, "On a Remarkable Bacteriolytic Element Found in Tissues and Secretions," a research paper wherein Fleming establishes his discovery of lysozyme and foreshadows his accidental discovery of penicillin. In this paper, Fleming observes how a specimen of nasal mucus prevents the growth of bacteria on experimental agar plates, "with the exception of the occasional staphylococcus colony."[7] Fleming then researches the action of tears, saliva, cartilage, tissue, and more on the same bacterial colonies, each with similar results. It is at this point Fleming surmises the presence of lysozyme, so named for its ability to "lyse," or tear apart the cell walls of bacteria. Fleming also found a large amount of lysozyme present in egg white, an interesting discovery in and of itself, as a 1909 research paper by early biochemist Laschtschenko showed egg white to have antibacterial properties.[8] Lysozyme from the samples Fleming took of tears and mucus mixed with saliva successfully killed both staphylococci and streptococci bacteria, albeit in small amounts, but bolstered Fleming's goal of finding antibacterial agents for malicious bacteria. In doing so, Fleming showed lysozyme to be on the front line of the human's immune system, acting in vulnerable spaces, particularly our eyes and nose, to kill bacteria.

The Discovery of a Lifetime

Fleming continued his pursuit of antibacterial agents over the next several years, looking for more potent ones at every turn. In 1928, his pursuit would pay off. Prior to going on vacation at his country home, a modest house coined "The Dhoon" in the Barton Mills village of Suffolk, Fleming pressed forward with work, inoculating several agar plates with the bacterial strains of staphylococcus variants.[9] Fleming left them on the benchtop, open to the air, and on returning to the laboratory from his family vacation, he noticed an unusual mold on one of the plates, with the staphylococci colonies near the fungi destroyed and others taking up residence away from the fungi along the far reaches of the agar plate. Fleming immediately went on a quest to search out the properties of this mold, which he first noted as *Penicillium rubrum* in the now pivotal 1929 paper "On the Antibacterial Action of Cultures of a Penicillium, with Special Reference to Their Use in the Isolation of B. influenzæ" reporting his findings on penicillin (it is of note that Fleming's original *P. rubrum* is now known to be *Penicillium rubens*, but out of deference to the reader, I will refer to the mold as the genus name *Penicillium* throughout). In the paper, many references are made to the color of the mold, which starts as white before often becoming dark green, then black, and then yellow in color, a spectrum of colors that may have aided Fleming in initially spotting the mold.[10] Fleming tested several other molds for antibacterial properties, but the *Penicillium* mold was the only one successful in killing bacteria. On the agar plates where the *Penicillium* mold was placed, streaked colonies of staphylococcus, streptococci, pneumococci, and gonococci (the bacteria responsible for gonorrhea) all retreated from the penicillin, with the streptococci, pneumococci, and gonococci colonies all turning clear in an evident show of their lysis and thus destruction. Due to the retreating bacterial colonies, Fleming had a hunch that it was a substance secreted by the *Penicillium* mold exhibiting the antibiotic activity. Fleming showed himself to be correct through a number of careful dilution experiments in which the substance from the mold was still active in fighting bacteria when diluted by a factor of 800, and he named the substance penicillin.[11]

In his 1945 Nobel Prize lecture, Fleming notes the simple manner in which he named penicillin. Instead of naming it for himself or something of standing, he took the route of naming it after that from which it is derived, specifically following the example of the naming of digitalin from the plant *Digitalis*. In another breath of humility in this speech, Fleming also conjectured that he could have written the original penicillin paper from the point of view of having hypothesized the entire interaction after sincere study and thought and appeared much smarter. Instead, Fleming noted, "My only merit is that I did not neglect the observation and pursued the subject as a bacteriologist."[12]

In addition to its action on bacteria, Fleming showed penicillin to be non-toxic within the 1929 paper through a variety of ways. He injected it into rabbits and mice as well as flowing a broth containing penicillin (Fleming was unable to isolate penicillin, so this was the best he could do) over an open, infected wound on a man and irrigating the eye of another every hour for a day with no detrimental effect. Penicillin also did not interfere with leukocytes as flavine did. This lack of toxicity, combined with its successful antibacterial properties in the presence of multiple harmful bacteria, including staphylococci, gonococci, streptococci, and pneumococci, set the groundwork for penicillin's eventual use as an antibiotic.[13]

Historically, it has been thought that the fungi Fleming observed floated in through an open window into the lab before settling on one of the staphylococci-inoculated plates. This is unlikely the case, as the windows in the building were difficult to reach and not often open. Instead, it is more likely that in yet another incidence of sheer luck in the discovery of penicillin, the *Penicillium* spores floated into Fleming's lab from the downstairs lab of Dr. C. J. La Touche, who was working to decipher a correlation between allergies and fungi. La Touche likely would have had a sample of *Penicillium*, as La Touche supplied Fleming with the other molds he tested for antibacterial properties in his 1929 paper.[14]

Despite his findings, Fleming's 1929 paper on the discovery of penicillin was not well received. Fleming's own inability to isolate penicillin, a development that would not occur until more than a decade later, stalled further work. Fleming did not lose hope in others, however, as he kept showing the discovery to people and giving away samples of the fungi so they could try to isolate penicillin themselves. Fleming even gave away sterilized samples of the mold on paper discs at his Nobel lecture.[15]

Although Fleming did not succeed in isolating penicillin, he did take part in one of its early applications as a therapeutic. In 1930, one of Fleming's former students at St. Mary's Hospital Medical School, Cecil George Paine, while a member of Sheffield University's Department of Pathology, used a crudely filtered mold broth given to him by Alexander Fleming in what stands as the first medical application of penicillin in humans.[16] Under Paine's care were two infants suffering from gonococcal conjunctivitis exhibiting significant discharge since birth. Paine dropped the mold broth into their eyes and cured both infants.[17] Dr. Paine also used penicillin in 1931, when a miner with a shard of stone in his right eye came under his purview. The eye was infected with pneumococci, and within forty-eight hours after treatment with the penicillin filtrate, the infection was gone. Paine, like Fleming, had his own difficulties in championing the use of penicillin, with no publications written by Paine or lectures given (that we know of) to highlight its curative effects.[18]

Passing the Torch

Fleming's paper on the discovery of penicillin would lie dormant for nearly a decade until 1938, when Ernst Boris Chain of Oxford University came across the paper and insisted his laboratory boss Howard Florey read the article. Howard Florey then turned the might of his research group, which included Norman Heatley and Edward Abraham along with Chain, onto the problem of purifying penicillin from *Penicillium* mold. The true work on turning penicillin into a drug now began.

The research group had prior exposure to Fleming's work through Edward Abraham, a structural wizard who was the first to crystalize lysozyme in 1937.[19] In 1942, Abraham would become the first to posit the correct structure of penicillin and its ever-important beta-lactam ring, which is pivotal to the molecule's antibiotic properties.[20] Ernst Boris Chain, a biochemist, would be the enigma and workhorse of the group, solving Fleming's conundrum by isolating penicillin from *Penicillium*. Chain's path to Florey's laboratory is particularly circuitous, as he escaped Berlin right as the Nazis rose to power, with his mother and sister dying in the Holocaust. Chain played a pivotal role in isolating penicillin by varying the pH of the growth conditions of *Penicillium* and then altering the temperatures to extremes before recovering a fingernail-sized amount of purified penicillin from gallons of mold broth.[21] Chain succeeded where no one else was able, not even Fleming himself despite many attempts and the hearty urging of others to attempt to isolate penicillin. Ernst Boris Chain also discovered that penicillin did not exist as an enzyme as Fleming's lysozyme did but that penicillin existed as a small molecule.[22]

Of Florey's cadre, it would be Norman Heatley who would play the role of jack-of-all-trades, growing *Penicillium* mold broth in bedpans and anything he could get his hands on. Heatley later perfected the process by which Florey's Oxford University group would extract and isolate penicillin.[23] Ever protective of his work, Heatley personally kept watch over the first animal experiments making use of their isolated penicillin, staying overnight during a weekend in May 1940 to observe the fate of eight mice injected with a lethal dose of hemolytic streptococci. Four of the eight received doses of penicillin, and these four were the sole survivors.[24] A repeat of this experiment showed Heatley and Florey what they hoped: that their isolation and therapeutic application of penicillin worked. The work carried out here would be published in 1940 as the paper "Penicillin as a Chemotherapeutic Agent" in the world-renowned medical journal *The Lancet*, staking the claim of Oxford University's Florey group to penicillin and giving new hope for turning the molecule into a drug.[25]

In September 1940, an opportunity for Florey, Chain, and the rest of the Oxford University group arose to show the might of penicillin. A local policeman, Albert Alexander, scratched his mouth while working in his rose garden. The scratch quickly became infected and spread to his eyes and the rest of his face. In Oxford's Radcliffe Infirmary, Alexander received sulfa drugs as an antibacterial treatment, but this was unsuccessful, as abscesses developed on his eyes, lungs, and shoulder.[26] It is important to note that a story contradicting the rose garden mishap exists, one a little more believable given the injuries and Albert Alexander's line of work. In this recollection of the incident, Albert Alexander is on loan to a Southampton police station due to a wartime mutual aid agreement between precincts when the Southampton station is bombed, leading to Alexander's injuries.[27] Regardless of the route Albert Alexander took to his bad luck, Florey and Chain heard of his plight and championed penicillin as the cure for Alexander. After undergoing five days of injections, Albert Alexander began to recover. Unfortunately, Alexander took a turn for the worse, and Florey and Chain lacked enough purified penicillin to completely remove the infection. Albert Alexander died soon thereafter.

At the time, an astounding 2,000 liters of *Penicillium* mold culture fluid was needed to provide enough purified penicillin to carry out a single person's treatment.[28] The vast size of the problem before them provoked the Oxford research group to reach out for help. In 1941, Howard Florey and Norman Heatley departed to the United States to ask industrial pharmaceutical groups for help in growing enough *Penicillium* mold broth to isolate significant quantities of purified penicillin. A smidgen of paranoia waded in as Heatley and Florey began to be concerned their sample vial of *Penicillium* might be stolen or lost. After all, England was on a wartime footing. Heatley eventually convinced Florey to smear their coats with the mold to ensure they would have access to a sample. During the trip, Florey and Heatley enlisted the directors of the U.S. Department of Agriculture's Northern Regional Research Laboratory, with Heatley staying behind for six months to aid them with the growth of *Penicillium* while Florey traveled east to garner the interest of the U.S. government and private pharmaceutical companies.[29] Thankfully, the duo was quite successful in their journey to the United States.

Although the Albert Alexander ordeal started with hope and ended in disaster, opportunities continued to arise for the application of penicillin in humans. In March 1942, Anne Miller suffered a miscarriage, and a subsequent streptococcal infection of her blood left the young woman delirious and hospitalized for a month in Connecticut's New Haven Hospital. During this time, Miller routinely experienced a fever hovering around 107 degrees Fahrenheit. Hearing of the experimental drug penicillin, Miller's doctors obtained a small amount from a Merck & Co. laboratory in New Jersey and injected their ailing patient with it. Overnight, Anne Miller's temperature dropped, and she was no longer delirious. Miller is viewed by history as the first person whose life was saved by the use of penicillin, and she would go on to live to the age of ninety before passing away in 1999.[30] All of this was thanks to a tablespoon of penicillin, a tablespoon that amounted to half the entire supply within the United States at the time.[31]

As luck (or unluckiness, depending on your point of view) would have it, around this time, researchers tackling the penicillin supply problem learned that penicillin rapidly left the body through the urine and in very large amounts. An early paper on the topic noted the excretion of penicillin within four hours was as high as 40 to 90 percent of the injected dose.[32] At such a high concentration, penicillin could easily be extracted from the crystalized urine of a patient and recycled to be used in another patient—gross, yes, but a boon to a world where a tablespoon is half a country's supply of the drug. The recoverability of penicillin from urine was likely known at the time of Anne Miller's administration of the drug, as an anecdote details Dr. Heatley of Oxford University delivering vials of penicillin and carting away gallons of urine back to New Jersey and the Merck laboratory.[33]

The first wide-scale application of penicillin would come later in the year after the tragic November 28, 1942, Cocoanut Grove nightclub fire in Boston. A busboy lit a match to see better for lack of a flashlight while changing a light-bulb, and within minutes, flames enshrouded the building, killing 492 people. Physicians were in search of anything to aid those who survived, with Merck & Co.—unable to supply enough pure penicillin at the time but wanting to be of aid and support—sending thirty-two liters of *Penicillium* mold broth via police escort for those at Massachusetts General Hospital with the hope of stemming some of the staphylococcal infections to which burn victims are susceptible.[34] This treatment echoes the 1930 work of Cecil George Paine using a crudely filtered mold broth to treat gonococcal conjunctivitis in infants.

In 1943, amidst the backdrop of World War II and looking for anything to prevent the Axis Powers from gaining an advantage, the British and U.S. governments placed a ban on the publication of any research pertaining to penicillin. At the same time, the two governments set up a series of channels to exchange information at monthly intervals between U.K. and U.S. research groups—both academic and industrial—working to get penicillin to the battlefield. Included in the industrial category are the companies that would go on to become the pharmaceutical giants Merck, Bristol Meyers Squibb, Pfizer, and Eli Lilly.[35]

Despite the best efforts of scientists in the United States (spurred on by the earlier visit of Florey and Heatley) and the United Kingdom, growth of and purification of penicillin was painfully slow and yielded disappointing results until a serendipitous discovery at a Peoria, Illinois, farmer's market. Peoria already stood out as a hotbed of penicillin research due to the presence of the U.S. Department of Agriculture's Northern Regional Research Laboratory, wherein a group was tasked with developing industrial-sized scale-up for the growth and purification of penicillin.[36] This was common knowledge to the people of Peoria, as it was routine for the lab to receive samples of moldy fruit from concerned citizens wanting to do their part.[37] On a fateful day in 1943, Mary Hunt, a laboratory assistant, viewed a cantaloupe with a "pretty, golden mold" on it at a Peoria fruit stand. The mold turned out to be a new species of *Penicillium* that yielded substantially more penicillin and even more so after scientists subjected the mold strain to additional mutations by bombarding it with x-rays until it yielded a 1,000-fold increase in penicillin.[38]

How Penicillin Works

With the mystery of penicillin's isolation and manufacturing nearly solved, now would be a good time to elucidate how penicillin works to kill bacteria. Penicillin is a relatively small molecule; the form of penicillin they were purifying comes in at 334 daltons, well in line with those pharmaceutical companies might look for in the course of modern drug discovery. Key to penicillin, however, is a four-membered beta-lactam ring. This quirk of chemistry allows penicillin to bind to and thwart the action of an enzyme in bacteria known as DD-transpeptidase. DD-transpeptidase works to create peptidoglycan bonds between sugars and amino acids and in doing so maintain and strengthen bacterial cell walls as bacteria divide.[39] DD-transpeptidase now belongs to a group of enzymes known as penicillin binding proteins. The cell walls of bacteria open up slightly as they are about to divide, with enzymes like DD-transpeptidase needed to help fill in this gap with more peptidoglycans. Without the action of these enzymes, the cell walls are overcome with osmotic pressure from outside the cell and water rushes in, causing the bacteria to essentially burst and die.[40] Penicillin affects gram-positive bacteria but does not affect negative bacteria, as gram-negative bacteria have an extra layer consisting of lipopolysaccharides that protects the peptidoglycan cell wall beneath. The lipopolysaccharide layer is made of lipids (think fats and oils) bound to sugars, and the enzymes penicillin act on have nothing to do with strengthening this lipopolysaccharide layer. Ernst Chain and Edward Abraham discovered bacteria could produce an enzyme capable of destroying penicillin shortly after isolating the drug in 1940.[41] This resistance to penicillin occurs when bacteria develop beta-lactamases, enzymes that destroy the functionally important beta-lactam ring on penicillin and render it unable to bind and carry on its action of inhibiting bacterial cell wall growth. These beta-lactamases render penicillin useless for therapeutic treatment if the bacteria in your body develop them.

Ramping Up Production

With the chance discovery of a moldy cantaloupe, the United States and United Kingdom were now able to manufacture the quantities of penicillin desired after years of toiling with tablespoon-sized amounts and the recycling of urine. The U.S. War Production Board stepped in and set a goal of 2.3 million doses of penicillin to be ready prior to implementing Operation Overlord, the June 1944 invasion of Normandy. Twenty-one factories in the United States were dedicated to around-the-clock growth and purification of penicillin. Howard Florey took note of the U.S. manufacturing might in 1949, saying, "Too high a tribute cannot be paid to the enterprise and energy with which the American manufacturing firms tackled the large-scale production of the drug. Had it not been for their efforts there would certainly not have been sufficient penicillin by D-Day in Normandy in 1944 to treat all severe casualties, both British and American."[42] By January 1945, the United States was capable of supplying 4 million doses of penicillin.[43] The form of penicillin they isolated is penicillin G, now known as benzylpenicillin, as more variants of penicillin were discovered and created over time. By March 15, 1945, injectable penicillin would become available to the general public and for sale in pharmacies across the United States.[44] Those in the United Kingdom would have to wait a little bit longer, with penicillin released in the country on June 1, 1946.[45] No one received a patent for penicillin, to Ernst Chain's chagrin, as Howard Florey (and possibly Alexander Fleming) viewed it as unethical to pursue a hold on such a lifesaving pharmaceutical at any time, let alone amidst the backdrop of World War II. Controversy arose as to whether a patent on penicillin could even be attained due to its existence as a natural product of a mold. Eventually, the process to create it received patent protection but not the molecule itself.[46]

As World War II ended, penicillin research continued. Edward Abraham's keen supposition of the structure of penicillin would be confirmed by Dorothy Crowfoot Hodgkin, as she solved the crystal structure for penicillin as people literally celebrated Victory in Europe Day in the streets right outside of her laboratory door.[47] An oral version of penicillin, penicillin V would become available in the early 1950s after some concern over the patent rights to this modified form of penicillin. Named so by the researchers for the German word for "confidential" (and unfortunately not for the better choice, at least story-wise, "V for Victory!"), penicillin V is also known by the chemical name phenoxymethylpenicillin.[48] Massachusetts Institute of Technology chemist John C. Sheehan would conduct the first total organic synthesis of a penicillin, the newly introduced penicillin V, in 1957, removing the necessity of giant vats of mold broth and fungi altogether.[49] Penicillin itself would be replaced in 1961 by an improved antibiotic, ampicillin, which would then be replaced by scores of others.

Dividing the Prize

Fleming, though painfully quiet and shy by most accounts, would receive more than his fair share of the accolades for discovering penicillin despite the herculean efforts of Ernst Boris Chain, Sir Howard Walter Florey, and the often-overlooked Norman Heatley and Edward Abraham in spearheading the wide-scale manufacturing of penicillin. Among these accolades is a 1944 *Time* magazine cover story featuring a stern painting of Alexander Fleming with the text below declaring, "Penicillin will save more lives than war spends."[50]

When it came time to hand out 1945's Nobel Prize in Physiology or Medicine for the discovery of penicillin, the Nobel committee split the accolades between Alexander Fleming, Howard Florey, and Ernst Boris Chain. The committee honored the accomplishments of Florey and Chain, proving that penicillin would never have reached the public without their work. Sadly, due to Nobel Prize rules limiting a single award to three individuals at a maximum, Edward Abraham and Norman Heatley did not take part in the official awarding of the prize. Their contributions are not forgotten, however, with Norman Heatley receiving at the age of eighty the first honorary doctorate in medicine awarded in the near millennia history of Oxford University.[51] Heatley's overlooked contribution would be lionized in time, with Florey's successor at the University of Oxford, Professor Sir Henry Harris, stating in a lecture honoring the accomplishments of Florey and his group, "Without Fleming, no Florey or Chain, without Chain no Florey, without Florey no Heatley, without Heatley no penicillin."[52] Heatley stayed forever humble with this refrain: "I was a third-rate scientist whose only merit was to be in the right place at the right time."[53]

In his Nobel Prize lecture, Fleming ends with a warning against underdosing penicillin. Overdosing did not persist as an issue due to penicillin's nontoxic effects on the majority of the population. Underdosing, however, either by taking too little of the antibiotic or by taking it for too short of a time period, posed dire consequences, as it gave bacteria a chance to develop resistance. In Fleming's "Mr. X" example, the symptoms of the patient disappear after he administers himself penicillin, but Mr. X does not take enough penicillin to kill the bacteria and instead develops a resistance to the drug. Mr. X later contracts a case of pneumonia and then dies due to bacterial resistance of penicillin. This hypothetical example puts on display the issues underlying self-administration of antibiotics, as Fleming ends with, "If you use penicillin, use enough."[54]

My personal favorite of all the awards received by the discoverers of penicillin is a quizzical but practical one. Bullfighters at the Las Ventas bullring in Madrid commissioned a bust of Alexander Fleming and a statue of a bullfighter taking his hat off in a sign of praise for the legendary scientist with a plaque saying, "To Dr. Fleming, with gratitude from bullfighters."[55] The reason? A large number of bullfighters believed their lives were saved due to the application of penicillin after being gored and trampled in the bullring.

So, as you can now tell, the story of penicillin reaching the shelves of our pharmacies and hospitals is a hard-fought one born out of chance—not just the chance that a spore of *Penicillium* floated down the hallway and up a stairwell into Fleming's lab and onto a bacteria-inoculated agar dish while he was away on vacation but a number of others as well. What if John Fleming never wills a portion of his estate to a young Alexander Fleming, who is left toiling away in a shipping office? What if an eager if not self-serving rifle team captain never points Fleming in the direction of Sir Almroth Wright and a career in research? What if Fleming never goes on his fateful vacation and instead keeps the agar dishes clean and is occupied by other experiments? What if Ernst Boris Chain never escapes Berlin after the rise of the Nazi Party? What if Mary Hunt never comes across a moldy cantaloupe in Peoria, Illinois? So many crucial break points in the pursuit of penicillin exist that it is nearly unbelievable that penicillin ever became a viable medication at all. In this tale of science, luck takes its share of the accolades, but luck never comes into the equation without the combined effort of Alexander Fleming, Howard Florey, Ernst Chain, Edward Abraham, Norman Heatley, and countless others whose hard work, diligence, and tenacity made penicillin a reality.

CHAPTER 2

Quinine

The structure of quinine. *Structure generated by author in ChemDoodle v 11.5.0, using information supplied by the National Institutes of Health*

The story of quinine is the story of malaria, as the pair have been synonymous with each other over the past two centuries. But this was not always the case, as the route to discovering and isolating quinine was hundreds of years in the making and touches nearly every continent. Let's take a look at the discovery of quinine as we try to tease fact from fiction in the story of this remarkable drug and along the way detail the desperate measures countries went to in acquiring and controlling the medication during the nineteenth and twentieth centuries.

Legendary Origins

Legend has it that a lost traveler suffering from a fever was stranded in the Andean jungles of the Peruvian Amazon rain forest and came across a pool of stagnant water. Thirsting for anything, he drank his fill. The traveler found the water to be bitter and soon became concerned he was poisoned, wondering if the "quina quina" trees surrounding the stagnant pool played the culprit. The opposite would turn out to be true, and within time, his fever went away. On returning to his home village, he shared his tale of the bitter water and the quina quina tree, leading the villagers to begin to use the bark of the quina quina to cure fevers.[1] Whether this legend rings true or not, the Quechua people of Peru are known to have used the bark of a tree to treat fevers and chills, and in time, the quina quina tree would become known by a different name.

Cinchona is the genus name for a plethora of shrubs and trees containing quinine but with some species containing more than others, as we will see. These evergreen trees are enigmatic, growing to a height of ten to thirty meters with clusters of white or pink flowers. Only the bark of the tree contains the much-sought-after drug quinine. The eighteenth-century Swedish botanist Carl Linnaeus is responsible for the name *Cinchona*, having named it in honor of the wife of the viceroy of Peru during the 1630s, the Spanish countess of Chinchon. By Linnaeus's understanding of a report recorded by Italian physician Sebastian Bado in 1663, the countess of Chinchon fell ill to malaria and was healed by the bark of a tree. As the legend goes, the countess was so amazed by the bark's properties that she ordered a supply of it for her to hand out personally to citizens in the area confounded by similar illnesses. This tale appears to be more fable than history, with a meticulous diary kept by the viceroy's secretary not once mentioning the countess undergoing treatment for malaria or the bark despite making numerous mentions of bloodletting treatments for the viceroy's bouts with malaria. It also appears from other records that the countess most often associated with the story, Ana de Osorio, died three years before her husband was appointed viceroy.[2] Despite the incorrect nature of the story by which it was named, Linnaeus's scientific name prevails to describe this wonderful tree steeped in quinine.

While it was likely not the countess of Chinchon who promoted the use of *Cinchona* bark, we do know that Jesuit missionaries played a large role in the transfer of the bark back to Europe and its promotion as a medicine. The Jesuit priests no doubt learned of the tree from the native peoples they were trying to convert—among them the Quechua people—and took note of the bark's usefulness. One Jesuit priest in particular, Agustino Salumbrino, operated a pharmacy for the poor in Lima, Peru, in the 1600s and is believed to have sent the bark to Rome in 1631 within the bags of Father Alonso Messia Venegas.[3] The Jesuit (and later cardinal of the Catholic Church) Juan de Lugo spoke highly of the bark in Spain in the seventeenth century as a treatment for fevers, with the healing outer layer of the *Cinchona* tree known by then as "Jesuit's bark."[4] The *Schedula Romana* of 1649 included an antimalarial recipe for Jesuit's bark believed to be perfected by Cardinal Juan de Lugo. The recipe called for two drachmas (a unit of mass based on the Roman drachma coin, with a drachma roughly three and a half grams) of bark to be ground into a fine powder and mixed with extremely hot, strong wine. The strength of the wine was no doubt noted to counteract the bitterness of the bark. The wine would then be ingested by the patient at least once a day.[5] In England, the symptoms associated with malaria were known as the agues, with Robert Talbor publishing *Pyretologia: A Rational Account of the Cause and Cures of Agues* in 1672 and describing a cure for the agues as a mixture of "four vegetables, whereof two are foreign and two are domestick [*sic*]" while cheekily warning against the use of Jesuit's bark.[6] Talbor received as much fame and reward as anyone thus far for the use of Jesuit's bark, as Talbor became the royal physician to Charles II shortly thereafter and was made a knight after curing Charles II of the agues in 1678. Talbor also ventured into the realms of French nobility, lending his cure to Louis XIV and his son when they fell ill.[7] For his efforts, Talbor received 2,000 gold crowns and a lifetime pension from Louis XIV, with Louis XIV also negotiating an agreement that the king could learn the secret to the cure after Talbor died. On Talbor's death in 1681, Louis XIV discovered the secret of Talbor's cure to consist of seven grams of rose leaves, a little more than two ounces of lemon juice, and Jesuit's bark mixed with wine.[8] A reference is made to Jesuit's bark in the third edition of the 1677 *London Pharmacopoeia*, where it is named *Cortex peruanus* (i.e., "Peruvian bark" in English), proving the wide-scale use or at least knowledge of a possible use of *Cinchona* bark in England at the time.[9] The Spanish held most of the world's reserves of Jesuit's bark, making it the desire of at least one pirate, Englishman buccaneer Basil Ringrose, who stole large amounts of the bark before he and his men were discovered and massacred by the Spanish in 1686.[10]

Malaria's Long History

The word "malaria" comes from a combination of "mal" and "air," meaning "bad air," suggesting the disease was spread through the air. Malaria is a disease that we now know is spread by female *Anopheles* mosquitoes. But why just the females? Because only the female of the species bites, taking part in blood meals as a method of nurturing its developing eggs.[11] Our understanding of the way malaria is spread is rather recent compared to the disease's long history—it was not until 1898 that Surgeon Major (and later Sir) Ronald Ross of the Indian

Medical Service proved the disease used mosquitoes as its mode of conveyance.[12] Ross would win a share of the 1902 Nobel Prize in Physiology or Medicine for this discovery. (The quick awarding of the Nobel Prize indicates the importance of his findings to the world at large at this moment in history.) *Anopheles* mosquitoes transmit malaria by mixing a *Plasmodium* parasite, typically *Plasmodium falciparum*, in their saliva with the victim's blood while feasting. Within ten days to four weeks later, the fever and flu-like symptoms of chills and sweats associated with malaria settle in while the body's red blood cells are under attack.[13] Often associated with malaria is the coming and going of fever symptoms, with a fever recurring every second day noted as "tertian" and one recurring every third to fourth day as "quartan." The cycle of recurrence is dependent on the *Plasmodium* parasite responsible for the infection, with *P. falciparum* often associated with tertian fevers.[14] Malaria is not contagious in the way we think of the flu or another virus being contagious: to transfer malaria between two people, an exchange of red blood cells must occur.[15] Despite this, there were still 228 million cases of malaria in 2018, with 405,000 dying because of the disease. Most diseases and deaths occur in Africa, with Nigeria alone accounting for one-quarter of all cases worldwide.[16] The twentieth century saw between 150 million and 300 million people die from malaria, accounting for between 2 and 5 percent of all deaths in that century.[17]

Humankind's war against malaria dates back thousands of years, with references to tertian and quartan fevers dating back to the fourth and fifth centuries BCE in Greece and possibly earlier in India and China.[18] DNA evidence traces the presence of malarial parasites to Rome in 450 CE. Just north of Rome is the town of Lugnano, where archaeologists uncovered a graveyard with forty-seven children aged three and under. Inside the leg bones of one three-year-old girl buried within the cemetery, scientists found malarial DNA. The vast number of similar-aged children along with the ritual sacrifice of puppies (thought to be a ritual sacrifice, as the remains of the young dogs were discovered ripped to pieces and placed beside the corpses) found in the graveyard denotes the likelihood something terrible was killing the children. Researchers hypothesized that the deaths were the result of an outbreak of malaria based on the uncovered DNA evidence. Such outbreaks of malaria are likely to have been one of the many ongoing crises leading to the eventual fall of Rome.[19] Malaria would continue to be a problem over the next 1,000 years in many parts of the Eastern Hemisphere. It is hypothesized that malaria did not exist in South America or the rest of the proverbial New World until the voyage of Christopher Columbus in 1492, with his journey bringing the dreaded disease along with dreams of conquest.[20]

Malaria almost definitely played a role in proliferating the slave trade in the New World, as it was believed African slaves were not as susceptible to the disease as European colonizers or the indigenous peoples of the Americas. Charles

Mann puts forth this argument in his book *1493: Uncovering the New World Columbus Created*, stating, "Malaria did not cause slavery. Rather, it strengthened the economic case for it."[21] Mann goes on to say that the slave owners had no idea of the connection between malaria and the slave trade but that the slave trade spread as newcomers saw the economic success of those with slaves in lieu of indentured servants and sought to replicate such success. Mann concludes with an argument that slavery would exist independent of malaria, noting Massachusetts as an example of a state with a significant slave population and little or no malaria to speak of.[22]

Uncovering the Secret of *Cinchona*

The secret behind the fever-healing properties of *Cinchona* bark would be discovered in 1820, when a French toxicology professor, Pierre Joseph Pelletier, partnered with pharmacy student Joseph Bienaimé Caventou to isolate quinine—a yellow, sticky gum that would go into solution readily when in the presence of an acid or alcohol—from the bark.[23] Quinine is an alkaloid, the first-semester organic chemistry class name given to naturally occurring chemical compounds that contain at least one nitrogen atom. Alkaloids are soluble in ethanol, making sense of why early antimalarial recipes called for the powdered bark to be mixed with wine. Caffeine is another example of an alkaloid, with Pelletier and Caventou isolating the molecule responsible for most of our morning productivity for the first time in 1821.[24] With the isolation of quinine successfully carried out, proper standardized dosages could be administered to patients, and long gone were the days of ingesting copious amounts of dried bark in hot wine.

At 324 daltons, quinine is well within the size range of a small-molecule drug. Interestingly, we do not know the exact mechanism by which quinine works to fight malaria even though it marks what is historically thought of as the first chemical compound to be successfully used to combat an infectious disease.[25] We do know quinine kills *P. falciparum*, possibly by preventing the removal of iron-containing heme, which is left over after *P. falciparum* consumes the hemoglobin contained in red blood cells. The leftover iron is toxic in large quantities, and it is possible that the presence of quinine allows the concentration of iron to rise to lethal levels in the parasite's digestive vacuole.[26] Although quinine's use predates the wide-scale regulation of pharmaceuticals in the United States, it is important to point out here, as a reminder, that the U.S. Food and Drug Administration (FDA) requires only that a drug be shown to be safe (within reason) and effective. The FDA does not require a corporation to know the exact mechanism of action or how the medication works in the body before approval, only that it works.

The successful isolation of quinine and its then two-century track record of success also made it the target of what is likely one of the first successful commercial medicines in the United States. Physician and entrepreneur John S. Sappington of Missouri exported *Cinchona* bark for quinine to be used in "Dr. Sappington's Anti-Fever Pills." Sappington started production of the pills in 1832 and publicized the medicine to fight malaria along with scarlet fever and yellow fever through two stores he operated out of Napton and Arrow Rock, Missouri.[27] In 1835, he founded the company Sappington and Sons to take the pills nationwide, with the pills particularly successful in Arkansas, Mississippi, Texas, and Louisiana.[28]

In the decades that followed Pelletier and Caventou's extraction of quinine from *Cinchona* bark, quinine and three other alkaloids from *Cinchona* bark, quinidine, cinchonine, and cinchonidine, were part of one of the first clinical trials in history. Thirty-six hundred patients over a three-year period beginning in 1866 were given the medications, with all four medications showing a greater than 98 percent cure rate for fevers.[29]

In this time period, access to quinine could be the difference between winning and losing in combat due to the high attribution rate of soldiers from malaria. Quinine rations were routinely handed out to Union soldiers during the American Civil War, while the South fared much worse, smuggling quinine in through the intestines of slaughtered animals and the heads of dolls.[30] British and Indian soldiers operating in China during the 1860s took quinine prophylactically.[31] Quinine also played a role in European expansion into Africa, where the threat of malaria was so great that Sierra Leone came to be known as the "White Man's Grave."[32]

Seeing the popularity of quinine and *Cinchona* bark on the world stage, Peru took drastic steps and cut off exports of *Cinchona* in the late nineteenth century in hopes of maintaining control over the plant. Two men, Charles Ledger and his Bolivian partner Manuel Incra Mamani, would do more to undermine these moves than any country could hope for and in doing so set up the Dutch as the premier suppliers of quinine to the world. The duo smuggled seeds from the Peruvian/Bolivian border, seeds that, as luck would have it, belonged to a species of *Cinchona* with an exceptionally high concentration of quinine. How high? As much as 10 percent in some samples. Ledger and Mamani first tried to sell the seeds in 1865 to the British, who spurned the pair. The Dutch did not repeat the folly of the British, using the seeds to set up their own growing operation on Java, an island in then–Dutch controlled Indonesia. The superb *Cinchona* variety grown from the seeds would be named *Cinchona ledgeriana* in honor of Ledger. Manual Incra Mamani did not fare so well—for his role in smuggling and subsequent seed expeditions, Mamani was beaten to death by police in 1871.[33] Using *C. ledgeriana* as its supply backbone, the Dutch set up the Kina

Burea, the world's first pharmaceutical drug cartel. The Kina Bureau ruled the quinine market for decades, setting production quotas and establishing prices while pushing public health campaigns against malaria.[34]

Quinine and World War II

The first report of resistance of malaria to quinine did not come until 1910, nearly 300 years into its use. Even then, the resistance was often not a full resistance, with quinine still having some effect on the patient.[35]

In a move to fortify their hold on the Pacific during World War II, Japan took over the Indonesian island of Java and the Dutch factory Banfoengsche Kininefabriek, the world's largest producer of quinine in 1942. This left the Allies without a supply of quinine, a deadly conundrum for those fighting in the Pacific theater. Quinine was not the only antimalarial available at the time, however. Atabrine existed and was distributed to Allied troops during World War II, but its lack of efficacy when compared to quinine, along with its side effects, led soldiers to refuse it. Another reason stood out as to why soldiers routinely refused atabrine: rumors spread by Japanese propaganda outlets successfully convinced many a soldier that atabrine caused impotence. This bout of patient noncompliance wasted nearly all of the 3.5-billion-tablet supply the United States had on hand during the height of World War II. Malaria was no laughing matter to those fighting in the Pacific theater, with Army Air Corps projections based on hospital admissions theorizing that four out of ten U.S. troops would be hospitalized for malaria unless drastic measures were taken.[36] The dire situation led the United States to take steps to secure *Cinchona* bark and thus quinine via the Board of Economic Warfare's Cinchona Missions. It is argued that the Cinchona Missions were second only to the Manhattan Project in importance to U.S. World War II–era research and development opportunities.[37]

The Cinchona Missions consisted of two phases. The first was a set of agreements to purchase all bark with a minimum quinine content of 2 to 3 percent from Colombia, Peru, and Ecuador, three of the major nineteenth-century quinine-producing countries. The second phase consisted of a series of expeditions carried out with local help by U.S. botanists and foresters to find new sources of *Cinchona* in the Andes.[38] The U.S. botanists and foresters would scout an area, and once *Cinchona* was found, the bark of the trees was subjected to the Grahe test—a test for quinine content wherein a pink smoke appears if quinine is present.[39] If successful, the U.S. botanists and foresters would turn over the backbreaking work to local *cascarilleros* to harvest the bark (*cascara* is the Spanish word for "bark"). Coppicing, the cutting down of a tree to preserve its root system and allow the tree to regrow, was the main method of harvesting used by the *cascarilleros*. Coppicing was carried out in lieu of stripping the bark and leaving the tree to stand, which would kill the tree in time. Remember that the bark is the only part of the *Cinchona* tree of importance if you are interested in quinine, forcing the *cascarilleros* to tirelessly work to remove the bark from the felled trees. Their job was not done at this point, far from it. Once the bark was removed, the *cascarilleros* carried seventy-plus-pound loads over mountainous terrain to makeshift fire-fed dryers and large ovens the United States set up to dry the bark and decrease its mass by up to 75 percent for shipping. In all, the United States imported 30 million pounds of dried *Cinchona* bark as part of the Cinchona Missions.[40]

Quinine's Role in Fighting Other Diseases

Despite quinine's enduring connection to malaria, other uses for the medication have been found over the centuries. Quinine was the first drug used to treat skin lesions stemming from the autoimmune disease lupus, with this application coming in 1894. Within four years, salicylic acids were used in conjunction with quinine to create a much more effective therapy for lupus, and this combination persisted as a first-line treatment against the disease for six decades.[41] Quinine and other antimalarials have been used to treat rheumatoid arthritis since the mid-1950s, as antimalarials are shown to reduce the swelling and tenderness of joints.[42] Quinine is also used as a combination therapy in conjunction with clindamycin for patients with severe *Babesia* infections, a parasitic infection spread by ticks that attacks the red blood cells with symptoms very similar to malaria, so much so that it is often misdiagnosed as such.[43]

Quinine was available over the counter as a pain reliever for nighttime leg cramps and restless leg syndrome up until 1994, when the FDA banned the over-the-counter sale due to the number of adverse effects that can occur with self-administration of quinine. The FDA determined that the risks of quinine overdose outweighed any potential benefit.[44] Despite the over-the-counter ban, quinine is still available via prescription, and the FDA is currently seeking to curb the off-label prescribing of quinine for leg cramps with the same calculus in mind.[45]

Despite FDA concern over its use for nocturnal leg cramps, quinine is used, in small amounts, in many popular drinks across the globe. Modern tonic water contains small amounts of quinine—eighty-three parts per million or eighty-three milligrams per liter is allowed by the FDA in the United States—with the small molecule giving tonic water its bitter taste.[46] Historically, tonic water contained significantly more quinine than it does now. Quinine is also a constituent of several modern liqueurs, including the French Lillet and Dubonnet and the Italian Campari. Peruvians are also all in when it comes to drinking quinine, making use of its bitter flavor along with Andean purple corn in a pisco morado tonic.[47]

Quinine, Cinchonism, and Its Successor

Quinine has a very small therapeutic window—the difference between a safe dose and an overdose is quite small. "Cinchonism" is the term given to the aftereffects of a quinine overdose (or consuming too much *Cinchona* bark), which can lead to skin lesions, blurred vision, tinnitus, and abdominal issues, including vomiting and diarrhea in mild but classic cases. In severe cases, blindness can occur, although the amount taken in such a severe case coincides with a suicide attempt in which the individual took fifty 200-milligram tablets of quinine sulfate along with alcohol.[48] Cinchonism can also affect the cardiac health, as esteemed Dr. Tinsley Harrison and colleague Dr. Joseph Reeves found when they evaluated then governor of Alabama, James "Big Jim" Folsom, for an ongoing bout of tinnitus. Reeves observed a prolonged QT on Folsom's electrocardiogram, a finding associated with a heart rhythm that can cause a series of rapid heartbeats and fainting. On following up with Folsom, the doctors learned he partook of multiple gin and tonics a day. The quinine in the tonic water likely caused the governor's complaints, as after Folsom switched to bourbon, the tinnitus and prolonged QT disappeared.[49]

Quinine is also, in part, responsible for blackwater fever, which occurred frequently at the turn of the twentieth century in soldiers who took quinine to prevent malaria while living and working in malaria hotbeds such as colonial Africa and South Asia. The frightening disease causes red blood cells to rupture, releasing blood into the urine—the figurative "blackwater" behind the disease's name—which is often followed by renal failure. The British army switched from quinine to atabrine for prophylactic treatment of malaria in 1943, with cases of blackwater fever quickly ceasing, leaving quinine as the empirical culprit.[50]

In time, quinine would be replaced by a synthetic analogue, chloroquine. The organic synthesis of quinine is not impossible and was carried out as a proof of concept in 1944 amidst the quinine and *Cinchona* bark shortages associated with World War II.[51] The total organic synthesis takes twenty steps, and, as such, quinine is quite difficult and costly to synthesize.[52] The most effective way of obtaining quinine is still through extraction from the bark of *Cinchona* trees.

Chloroquine would not exhibit the staying power of quinine, the latter of which did not see signs of resistance until 1910. *Plasmodium falciparum* began showing resistance to chloroquine in Southeast Asia and South America during the 1950s and worldwide by the 1980s. Chloroquine remains in use where infections are caused by other variants of the *Plasmodium* parasite. Quinine still sees use as a second-line treatment against malaria or a first-line treatment in areas where resources are limited, particularly in Africa.[53]

Interestingly, two of quinine's successors, chloroquine and hydroxychloroquine, were being used in some parts of the world as an experimental clinical trial treatment for COVID-19, with results pointing to the ineffectiveness of the duo.[54] Unfortunately, this did not stop people from taking their medical care into their own hands, and on March 24, 2020, an Arizona couple ingested chloroquine they kept in their home to treat parasites that developed on their pet koi fish. Their hopes? Prophylactic treatment against COVID-19. The overdose of chloroquine led to a fatal heart attack in the husband, with his wife hospitalized.[55] In June 2020, the FDA issued a warning against the use of either chloroquine or hydroxychloroquine outside of a clinical trial setting due to the distinct possibility of serious alterations in heart rhythm along with kidney and liver failure that can arise.[56]

CHAPTER 3

Acetylsalicylic Acid

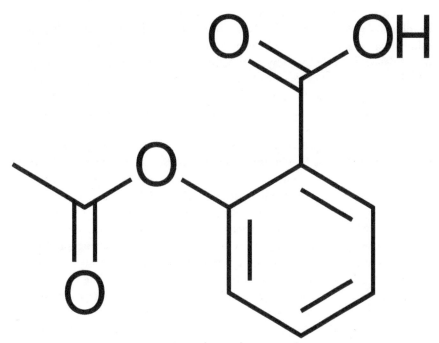

The structure of acetylsalicylic acid, better known as aspirin. *Structure generated by author in ChemDoodle v 11.5.0, using information supplied by the National Institutes of Health*

You no doubt know of acetylsalicylic acid by another name: aspirin. Early twentieth-century literati Franz Kafka is said to have described aspirin as one of a handful of things in the world that assuaged the pain of existence.[1] It is the most commonly used drug in the world, with at least 120 billion pills of aspirin made each year in China alone.[2] From the bark of the willow tree to its discovery and rediscovery decades later, aspirin plays a ubiquitous part of our lives and a daily part of many for the prevention of heart attacks and possibly even cancer.

An Ancient Use of the Willow Tree

The willow tree is part of the scientific genus *Salix*, a group of large trees with telltale weeping branches and narrow green leaves that hide within their bark a chemical that continues to change the world. A connection between the willow tree and medicine has existed for millennia. Clay tablets hewn by the Sumerians show they believed the willow tree to be vital to relieving pain and inflammation.[3] Babylonians used willow tree bark for pain and fever, while the Chinese made use of a particular species of the willow tree, *Salix babylonica*, to treat rheumatic pain and goiters. In Greece, Hippocrates wrote of using a brew of bark from the willow tree to treat ophthalmic pain as well as to ease pain in childbirth. Application of willow bark continued over the centuries, with Roman soldiers carrying large caches of the bark as they went into battle.[4] Interestingly, use of willow bark did not stay confined to the Eastern Hemisphere. Traces of salicylic acid, a chemical eventually isolated from the willow tree, was found on pottery dating to the sixth to seventh century CE just outside Denver, Colorado, suggesting Native Americans knew of its pain-relieving properties and made use of an extract of the bark.[5]

Another modern discovery shows the Egyptians used willow tree bark in a number of medicines. Edwin Smith purchased the Ebers Papyrus in Cairo in 1862, an astounding document measuring twenty meters long and written in an abridged form of hieroglyphics dating to 1500 BCE.[6] The papyrus received its name from its eventual owner Georg Ebers, a German Egyptologist as well as novelist, who purchased the ancient medical text in 1873 while visiting the city of Luxor, Egypt. The Ebers Papyrus makes numerous references to the use of Egypt's willow trees to cool and dry wounds and treat inflammation as well as broken bones. The text does not stop with the willow tree either, as the tome carries within treatments for ulcers and tumors as well as a recipe for treating excessive urination, a symptom of what we now know as diabetes. The remedy for excessive urination? Water from a specific pond, elderberry, dates, cucumber flowers, and milk.[7]

It is important to note that with so many varied civilizations across the globe making use of willow tree bark to heal, several of these were likely independent discoveries of its usefulness, especially in the case of the Colorado discovery. Our ancestors were far more advanced than we often give them credit for. Their lack of modern communication tools and instrumentation did not prevent them from examining a vast scientific world in their own right.

The willow tree's healing properties are due to high levels of salicylic acid in its bark, but it's not the only source of salicylic acid found in nature. Salicylic acid is also found in lower concentrations in a variety of plants where it acts as part of their defense mechanism, including the shrub *Spirae alba*, which is also known as meadowsweet. There are some unusual sources of salicylic acid used throughout history as well. Foremost of these is the castoreum of beavers, a sticky substance stored within sacs under the tail of the rodents and used to mark their territory. These sacs often rivaled the pelts of the beaver in value, with the castoreum becoming inundated with salicylic acid as the animal feasts on willow bark. Medieval physicians used the castoreum sacs to treat headaches, which would be appropriate, but they also employed castoreum in some not-so-useful ways as well, including treating impotence.[8]

Why is salicylic acid so prevalent in the willow tree and found to be present in so many other plants? Salicylic acid is a plant hormone, one defending the plant against pathogens by itself and with the help of salicylic acid derivatives. For example, when tobacco plants are inoculated with tobacco mosaic virus, the tobacco plant quickly turns salicylic acid into methyl salicylate, which can then spread from the tobacco plant to the air and the surrounding tobacco plants. It is hypothesized that methyl salicylate lands on the other plants and is then converted back to salicylic acid, which acts as a signal for the other plants to begin going into a defensive mode, turning on genes that code for aspects of protection and disease resistance.[9]

Leveraging the Willow Tree with the Power of Chemistry

Our modern use of the willow tree starts in eighteenth-century England with the Reverend Edward Stone. The good reverend used willow tree bark to treat fever and carried out his own empirical study to communicate its benefits, which he summarized in 1763's *An Account of the Success of the Bark of the Willow in the Cure of Agues*. This study was initiated after Stone noticed willow bark had a bitter taste similar to the revered (and much more expensive) Peruvian Jesuit's bark. On a hunch, Stone left a pound of willow bark near a baker's oven to dry for three months. He then crushed it into a powder and treated himself and others with the powdered bark whenever a fever arose. Stone claimed success in curing the fever of more than fifty people in the next five years using his supply of willow bark, but he did note that some quartan fevers—a common symptom of malaria—persisted. How did he cure such fevers? By mixing Jesuit's bark in a one-to-five-part mixture with willow tree bark, Stone was able to make these fevers go away as well.[10]

The greater expense of Jesuit's bark aside, there was another reason for the transition from *Cinchona* to willow tree bark in Edward Stone's corner of the world. During the early nineteenth century, much of England was subject to a blockade put in place by Napoleon Bonaparte, preventing precious Jesuit's bark from reaching the shores of England.[11] Willow tree was the best many could manage, even after quinine had been successfully extracted in 1820 from *Cinchona*, due to the inherent difficulty in synthesizing quinine.

Facing this embargo amidst the dawn of organic chemistry, Johann Buchner successfully isolated yellow crystals from willow bark and named it salicin, after *Salix*, the genus of the tree, in 1828.[12] A decade after Buchner successfully extracted salicin from willow tree bark, Raffaele Piria produced a more potent compound from the yellow crystals of the willow tree in 1839, naming it salicylic acid.[13] In 1853, French chemist Charles Gerhardt modified salicylic acid with the addition of an acetyl group (a carbon bonded twice to an oxygen, with the carbon also bound to a second carbon flanked by three hydrogens), creating acetylsalicylic acid, but the compound was unstable. Gerhardt would not be able to pursue this problem further, as he died three years later at the age of thirty-nine, with exposure to chemicals in the process of his work the likely culprit. Hugo von Gilm succeeded where Charles Gerhardt was unable, stabilizing the product and synthesizing a form of acetylsalicylic acid in 1859.[14]

In the four decades that followed, acetylsalicylic acid was essentially shelved, but its chemical predecessors did see some investigation into their therapeutic effects. In 1876, the first clinical trial of Johann Buchner's salicin was published. Thomas Maclagan of the Dundee Royal Infirmary in Scotland ingested salicin to test the dosage, then treated eight patients suffering from rheumatic fever with twelve grains (roughly 775 milligrams) of salicin every three hours. In each case, their fever vanished, but salicin was likely not widely accepted even after this study due to stomach issues that often arise after its ingestion.[15]

Bayer and the Creation of Aspirin

Twenty years after Maclagan's clinical trials, the story of acetylsalicylic acid begins in earnest when the German-based company Bayer AG steps in. Rising from the efforts of dye salesman Friedrich Bayer and master dyer Johann Friedrich Weskott, what we now know as the multinational pharmaceutical giant Bayer AG was founded on August 1, 1863. Prior to the 1880s, Bayer had essentially been a dye manufacturer specializing in creating synthetic dyes from coal tar derivatives, an important albeit less altruistic offshoot of the chemistry trade.[16] In time, Bayer would turn its eye to the pharmaceutical market, purchasing medicines invented by others and then footing the bill to market and mass-produce. Soon Bayer began to design drugs in their own laboratories, all the while still purchasing the worthy efforts of other inventors. One such inventor would be Arthur Eichengrün, who came to Bayer with protargol, a treatment for gonorrhea, and left with a position in the company on October 1, 1896.[17] Protargol, a mixture of silver and protein, is an astonishing drug in its own right. Protargol served as the frontline treatment against gonorrhea for five decades until penicillin was isolated and manufactured in large quantities in 1945.

With Eichengrün within the fold at Bayer AG, the company began seeking out derivatives of salicylic acid that would not produce stomach ailments. Felix Hoffman, who worked with Eichengrün in the fabled Room III of Bayer Laboratories, successfully rediscovered acetylsalicylic acid, synthesizing a pure form of the compound on August 10, 1897.[18] In later years, Hoffman would state that his interest in finding derivatives of salicylic acid stemmed from a desire to cure the ailments of his father. The elder Hoffman took salicylic acid for rheumatism, but the gastrointestinal side effects of the medicine made him vomit violently every time he did so.[19] After the synthesis of acetylsalicylic acid, Felix Hoffman immediately imbibed a small sample and noticed that it tasted differently than salicylic acid. This gave Hoffman hope, a hope that they succeeded in making a compound that was easier on the stomach.[20] Hoffman's success in the chemical arts did not end there, however. Less than two weeks later, Felix Hoffman also synthesized diacetylmorphine, better known as heroin, as part of Bayer's hopes to create a nonaddictive form of morphine.[21]

The head of Bayer's pharmacology division at the time, Heinrich Dreser, dismissed acetylsalicylic acid from clinical trials, believing the drug would be harmful to the heart. Eichengrün was not satisfied with Dreser's veto and instead set up his own clinical trials by spiriting samples of acetylsalicylic acid from the laboratory. Eichengrün recruited physician Felix Goodman and others in secret for clandestine trials as well as taking acetylsalicylic acid himself. Goodman and Eichengrün noted that the molecule quickly relieved pain and fever while not causing heart problems. Dreser continued to dismiss these results until a higher power at Bayer AG intervened in the form of Bayer head of research Carl Duisberg, who ordered more trials to validate Eichengrün's results.[22] These trials, as time tells us, were successful, and Bayer officially had a wonder drug on its hands.

With clinical trials performed and efficiency proven, what was left before bringing acetylsalicylic acid to market? Naming acetylsalicylic acid so as to distance it from its less palatable cousin, salicylic acid. The name "aspirin" would be given to the drug once in production, a combination of "a" from "acetyl," "spir" from the Latin *spiraea* for the meadowsweet plant (yet another plant from which salicylic acid can be obtained), along with the ending "in," which stands for nothing, as it was merely a popular ending for drug names at the time.[23] Although there existed concerns customers might confuse "aspirin" with the verb "aspirate," "aspirin" still won over the second-choice name for acetylsalicylic acid, "euspirin." In dismissing the "eu-" prefix, Eichengrün would note that it "is generally used for improved taste and odor." Eichengrün likely had the final say in choosing the name "aspirin," as evidenced by a January 23, 1899, memo in which he makes a note preferring "aspirin" to "euspirin," with the memo signed by Felix Hoffman, Carl Duisberg, and Heinrich Dreser and without any such notes from the trio.[24]

Dreser went on to accrue a fortune from the sales of aspirin due to an agreement he had with Bayer AG to receive royalties on any medicines that came out of his lab. Hoffman and Eichengrün received no such royalties for their discovery.[25] Hoffman would receive more than his fair share of the credit, however, as over the next several decades, Eichengrün's contribution in the discovery of aspirin would erode. This is due to Eichengrün's Jewish lineage and the rise of Nazism in Germany, where Bayer AG was helmed. Over the years, Eichengrün repeatedly insisted he fostered the idea for the creation of a salicylic acid derivative with fewer gastrointestinal side effects as well as formulating the synthesis for acetylsalicylic acid, with Hoffman acting as a soldier and merely carrying out Eichengrün's instructions.

Eichengrün left Bayer in 1908 to form his own chemical company in Berlin, a company specializing in the creation of flame-resistant materials using acetyl cellulose. During this period, Eichengrün also pioneered the field of injection-molded plastics.[26] Eichengrün's continued success after leaving Bayer also likely played a role in his separation from the aspirin creation narrative. It simply did not behoove Bayer to promote the efforts of someone no longer affiliated with them and running a rival chemical company. In 1934, a history of the discovery of aspirin was written in consultation with Bayer archives, giving significant praise and credit to Dreser and Hoffman. No such lauding of Eichengrün was within, with the scientist completely left out of the tale.[27]

Unfortunately, circumstances only grew worse for Eichengrün. His Jewish lineage caused him to lose control of his Berlin factory through a forced transfer in 1938. Eichengrün was married to an "Aryan" wife, with one scholar suggesting this was enough to keep him out of the concentration camps, at least until 1944, when the seventy-six-year-old Eichengrün was sent to the Theresienstadt concentration camp in the Protectorate of Bohemia and Moravia. Thankfully, he survived to see liberation at the hands of the Russians in 1945.[28] While interned at Theresienstadt, Eichengrün wrote a letter detailing his involvement in the creation of aspirin. The letter was later turned into a 1949 article in the journal *Pharmazie* that would be Eichengrün's swan song: his own telling of the history of aspirin.[29] Parts of the tale are heartbreaking, with Eichengrün recounting a visit to the Hall of Honour of a museum in Munich. The front of the museum was adorned with a large sign proclaiming that "non-Aryans" should not enter. Eichengrün went inside anyway and quickly saw a display for acetylsalicylic acid that championed it as a German achievement with credit given solely to Dreser and Hoffman. Later in the visit, he came across a similar display for acetyl cellulose, which Eichengrün popularized in his factory, with no inventors or contributors listed.[30] Most scholars now hold Eichengrün's account of the rediscovery of acetylsalicylic acid to reflect reality, with Eichengrün dictating directions to Hoffman, who carried out the synthesis.

Bringing Aspirin to Market, the Great Phenol Plot, and the Spanish Flu Boycotts

At just over 180 daltons, acetylsalicylic acid is a small molecule in the eyes of modern drug designers. The creation of acetylsalicylic acid is actually quite easy and is often a task performed by students in a first- or second-year undergraduate chemistry lab. The synthesis can be done in as little as three steps.[31] The lab is often very popular and memorable for students, as it provides a fruitful opportunity for novices to make use of organic chemistry techniques that are not terribly far removed from the world of drug design and synthesis while creating a final product, aspirin, with which the students are familiar. Heck, I might have considered using the aspirin I synthesized in my sophomore year to cure a headache later in the night.

In 1899, Bayer AG popularized aspirin by sending kits across Europe announcing its imminent release. At this time, aspirin was available only in a powder form and not the tablets to which we are accustomed. The mailing kits were sent to more than 30,000 physicians in what may be the pharmaceutical industry's first wide-scale "mail-out" of advertising information.[32] Bayer stood by a desire to make the drug available only via a physician's prescription to avoid any connection to less-than-reliable patent medicines that were prevalent at the time. Salicylic acid quickly fell out of favor for use against pain and inflammation once Bayer's wonder hit the market, but salicylic acid continues to see use to this day as a topical treatment for acne and warts.

Bayer's popular remedy was touted as a cure-all drug, with Alexei Nikolaevich, son of Russian Tsar Nicholas II, given the drug for his hemophilia in a time before aspirin's effect on thinning the blood was known. Grigori Rasputin ordered the "Western" drug to be thrown out prior to initiating his own response to Alexei's care, with the cessation of aspirin leading to an immediate improvement in Alexei's health regardless of the usefulness of any additional methods Rasputin tried. This improvement in Alexei's hemophilia symptoms led, in part, to Rasputin's acceptance within the inner circles of the court of St. Petersburg.[33]

In 1915, the popularity and wide-scale acceptance of aspirin led Bayer AG to manufacture the drug in a tablet form, and the company started selling it as an over-the-counter remedy.[34] At this time, World War I raged in Europe, with the United States yet to join the fray. Bayer, a German company with manufacturing interests in the United States—specifically, a factory in Rensselaer, New York, where they manufactured acetylsalicylic acid—was in a bind. The method of synthesis of acetylsalicylic acid relied on by the Rensselaer factory required large amounts of the chemical phenol to create the acetylsalicylic acid precursor salicylic acid. Phenol, however, was needed for a much different reason than pain relief at the time. Phenol suppliers in Great Britain, the number one source of phenol to the Rensselaer Bayer plant, were at the command of the British Parliament to use all of their phenol to produce the explosive trinitrophenol for use in war. The shortages of phenol almost led Bayer to shut down production of aspirin at the plant in April 1915.[35] Timing confounded this problem, as anti-German sentiments in the United States soon rose thanks to the May 7, 1915, sinking of the RMS *Lusitania* by a German U-boat, which killed nearly 1,200 people. The need for phenol on the part of Bayer still remained, with Bayer's conundrum coming through being a German company cut off from its American counterpart during a time of war, a counterpart that relied on a British export for its best-selling commodity. Fortunately for Bayer, a solution was at hand thanks to one of the more curious characters in U.S. science history, Thomas Edison, as he was also in need of large quantities of phenol. Why? To create discs to play on his line of phonographs. Unable to secure a steady supply of phenol, Edison ordered the construction of a plant to synthesize phenol from the precursor benzene. Edison's chemical plant was capable of producing twelve tons of phenol a day, three more tons of phenol a day than Edison required for the production of his early records. What did Edison decide to do with the rest? Sell it on the open market through a broker at the American Oil and Supply Company as prices for phenol were skyrocketing.[36] Edison quickly found a home for the excess phenol through the manipulation of a pair of German nationalists—Hugo Schweitzer, a German spy and industrial chemist, and Heinrich Albert, a German Interior Ministry official. The duo used German government funds to secure Edison's excess phenol before selling it to the American counterpart of German chemical firm Chemische Fabrik von Heyden, which then converted the phenol into salicylic acid and sold it to the American end of Bayer. The salicylic acid was then used to synthesize acetylsalicylic acid to keep the flow of aspirin going in the United States.[37] The benefit to a wartime Germany—apart from aiding a German-based company—may appear thinly veiled at this point. However, every drop of phenol put into the production of aspirin was a drop that the United States could not buy from Edison and use to manufacture explosives that could one day be used against Germany (the United States would not formally enter World War I until April 6, 1917, but after the attack on the RMS *Lusitania*, the United States firmly supported British interests in the war). Albert and Schweitzer's grift did not last long—a little less than two months—as a Secret Service agent tailing Heinrich Albert on July 24, 1915, stole his briefcase, one filled to the brim with compromising papers. The United States leaked the papers to the anti-German *New York World*, which published an exposé detailing the Great Phenol Plot on August 15, 1915.[38] Both Bayer and Edison were shamed, with Thomas Edison vowing to sell his extra phenol to the U.S. government.[39]

Within three years, Bayer would be embroiled in yet another controversy in the United States. When the Spanish flu took hold in the United States in August 1918, rumors swirled about its origins. Within the October 18, 1918, edition of the *Birmingham News*, Bayer took out an advertisement stating that "the manufacture of Bayer tablets and capsules of aspirin is completely under American control . . . being operated as a 100% American concern."[40] Why take out such an obtuse advertisement in a large, regional newspaper? To combat bizarre rumors suggesting the German-based Bayer company intentionally put the Spanish flu in aspirin as a part of a large-scale World War I operation. With the United States firmly entrenched in World War I by 1918, anti-German sentiment was high, and boycotts were common as even Bach and Beethoven were shunned.[41] Any boycotts of Bayer would be meaningless, however, as the United States took control of Bayer's U.S. assets on January 10, 1918, as part of a plan enacted to seize German assets just after the United States formally entered World War I, placing the corporation truly under "American control" as the advertisement stated.[42]

Aspirin was employed to treat the Spanish flu due to its pain and fever-relieving qualities. The U.S. surgeon general at the time, Rupert Blue, called for the use of both quinine and aspirin to ameliorate the symptoms of (but not cure) the Spanish flu, giving particular praise to the use of aspirin in a September 1918 statement. Deaths skyrocketed after Blue made this announcement, as October 1918 was the single month with the highest rate of mortality thus far from the pandemic.[43] Was there a connection? Possibly. There is evidence aspirin was being given to patients at levels we would now deem toxic, with dosages ranging from eight to thirty grams a day to fight the effects of the Spanish flu. Such intense aspirin therapy leads to a buildup of the molecule within the body, eventually leading to salicylate intoxication and its manifestation as pulmonary edema in some cases. Postmortem examinations of the lungs of many who died from the Spanish flu during this period note "wet" lungs, an observation suggesting pulmonary edema via aspirin administration played at least some role in the deaths.[44] While aspirin did ease the symptoms of the Spanish flu, it is quite likely it played a role in increasing the death toll of influenza when taken at what we now know to be toxic doses.

The end of World War I saw Bayer lose control of its U.S. trademarks as a part of war reparations. The bulk of Bayer's U.S. holdings were sold to Sterling Drug, Inc., at a no doubt embarrassing auction on the footsteps of Bayer's

Rensselaer, New York, factory for the sum of $5.3 million on December 12, 1918.[45] Sterling Drug would come under the control of the Eastman Kodak Company in 1988, which sold the worldwide Sterling Drug over-the-counter business operations to SmithKline Beecham in 1994. Bayer, ever on the hunt to gain control of its trademarks in the United States again, used this opportunity to buy the U.S. rights to what was Sterling Drug from SmithKline Beecham weeks later for the sum of $1 billion.[46] Do not feel bad for Bayer, however, as the company's aspirin sales alone account for more than $1 billion in sales a year.[47]

Aspirin's Second Life

Aspirin is a nonsteroidal anti-inflammatory drug, a class better known by its abbreviation NSAID. While aspirin was less bothersome to the stomach than its precursor salicylic acid, stomach irritation and bleeding remained an issue with patients. As with most drugs, innovation presses forward looking for new and improved versions, and aspirin would be no different. Aspirin would not see its first real challenger to the throne until the arrival of acetaminophen on the marketplace in the 1950s, which was soon followed by the NSAID ibuprofen in the 1960s. Both acetaminophen and ibuprofen gave patients less gastrointestinal issues while still fighting off pain and fever, and because of this, aspirin's use soon waned.

Aspirin's saving grace would be decades in the making. Enter now into acetylsalicylic acid's history Lawrence Craven, a Los Angeles family doctor who prescribed chewing gum mixed with aspirin to treat pain after tonsillectomies. Craven noticed severe bleeding issues arose if the patients used too much of the compounded chewing gum and posited to several obscure journals, including the *Journal of Insurance Medicine* and the *Mississippi Valley Medical Journal*, that aspirin might prevent blood clots and heart attacks.[48] It is not Craven's fault that he resorted to off-the-beaten-path journals for his research; he was simply a victim of being outside the traditional research path and lacked the academic capital needed to publish in higher-profile journals. Craven also lacked something very important in research: a control group of patients in his practice not taking aspirin to measure his findings against (this was understandable, as Craven was a family physician aiming to aid all of his patients if it was within his ability). Craven's first letter concerning his observations about aspirin was written in 1948 to the *Annals of Western Medicine and Surgery*, and in it, he communicated that of 400 patients in his family practice he had prescribed a single daily aspirin to over two years, none suffered a heart attack. To back his argument that aspirin stopped blood from clotting, Craven experimented on himself by taking twelve tablets of aspirin a day for five days to the point where he developed nosebleeds.

Craven continued to expand his observations, and by his third paper, he was now prescribing aspirin to men between the ages of forty-five and sixty-five who were overweight and led a sedentary lifestyle. In his final paper, Craven claims to have prescribed a daily prophylactic aspirin regimen to 8,000 patients, with only nine dying of heart attacks but with the subsequent autopsies of each of the nine revealing their death to be due to a ruptured aortic aneurysm.[49] The next year, 1957, Craven died of a heart attack at the age of seventy-six, possibly casting aspersions on this insistence that aspirin could ease heart conditions. Regardless of his methods and his setting in a family practice far removed from a clinical trial environment, Craven appeared to be on to something.

A study of aspirin's antiplatelet activity was carried out in 1967 with a control group, the very group Craven's studies lacked. This study examined the amount of bleeding and total clotting time, with the ten aspirin-taking subjects requiring more time for coagulation than the control group.[50] Although the result would come ten years after his death, Craven would be shown to be right. In 1988, a massive clinical trial with more than 17,000 participants who suffered a heart attack showed aspirin to reduce the risk of a second heart attack in the next five weeks by 20 percent. This result was quickly followed by trials that showed aspirin to be useful in the prevention of strokes.[51]

How Aspirin Works

It took nearly a century from Hoffman's synthesis of acetylsalicylic acid for scientists to discover the route by which aspirin relieved pain and inflammation in the body. In 1971, John Vane and his graduate research student Priscilla Piper showed aspirin to stop the formation of prostaglandins. Prostaglandins are lipids that act like hormones in the body and signal a cellular response—for example, the prostaglandin PGE_2 sends signals to the body that result in pain being felt. When aspirin inhibits the synthesis of PGE_2, these signals are never sent, and no pain is suffered by the patient.[52] For his work on the connection between prostaglandins and aspirin, Vane received a share of the 1982 Nobel Prize in Physiology or Medicine. One of the other two recipients, Sune K. Bergström, had proved the existence of prostaglandins three decades prior, providing the basis for Vane and Piper's work.

A few years after Vane's discoveries, the underlying reason for the inhibition of prostaglandins would be learned. In 1976, the enzyme cyclooxygenase became forever linked to aspirin, specifically COX-1, with aspirin shown to irreversibly bind to COX-1. This binding prevented COX-1 from transforming arachidonic acid in the body to a number of prostaglandins. Included among these prostaglandins is thromboxane A_2, a molecule that causes platelet aggregation.

With COX-1 irreversibly bound by aspirin, thromboxane A$_2$ cannot be synthesized by platelets during the remainder of the platelet's lifetime, leading to the decrease in blood clotting (and thus a decrease in heart attacks) seen when people take low doses (eighty-one milligrams or less a day) of aspirin.[53] COX-1 is also responsible for the synthesis of prostaglandins, which protect the gastrointestinal tract, explaining why the use of aspirin often causes stomach issues.[54] In time, a second cyclooxygenase would be discovered, COX-2. Aspirin binds to COX-2 but not as strongly as it does COX-1, with COX-2 responsible for prostaglandins leading to an inflammatory response in the body. Higher doses of aspirin are needed to see an observable decrease in inflammation via binding of COX-2.[55] There have been attempts to create drugs that only inhibit COX-2 to forgo the gastrointestinal issues associated with blocking the activity of COX-1. The most controversial of the COX-2 selective inhibitors is Merck & Co.'s Vioxx (generic name rofecoxib), which gained FDA approval in 1999 but was voluntarily removed from the market in September 2004. Why? Evidence showed that Vioxx led to an increased risk of heart attack and stroke in patients. According to Dr. David Graham, associate director of the FDA's Office of Drug Safety, between 88,000 and 139,000 Americans suffered additional heart attacks or strokes while taking Vioxx.[56]

Throughout the decades, the work of Vane and a legion of others would be put to the test in a series of studies and clinical trials. By 1994, the *British Medical Journal* would summarize these as the "Aspirin Papers" and cement the widescale use of aspirin for the prevention of heart disease as well as ischemic stroke (a type of stroke that occurs when an artery that supplies blood and thus life-giving oxygen to the brain becomes blocked). Additionally, it would argue for the acute use of aspirin during a heart attack.[57] Aspirin's second life was even more vibrant and moneymaking for manufacturers than the first, with daily regimens needed to stave off life-threatening events yielding a larger financial windfall than the every-once-in-a-while use of the drug to make a headache go away.

In time, aspirin could possibly be shown to be useful in preventing the onset of cancer as well. Now is also probably a good time to say that nothing in this book should be taken as medical advice, as many of the fledgling uses we will see for the medications within are not entirely vetted, nor do guidelines exist for all of their possible applications. If there is any utility, the mechanism for aspirin's fight against cancer likely revolves around COX-2. The activity of the enzyme COX-2 is observed to be increased in tumors, leading to the hypothesis that aspirin inactivating COX-2 decreases the proliferation of tumor cells and may decrease their resistance to apoptosis (programmed cell death).[58] One recent large study of aspirin's preventive use against cancer showed that taking any dose of aspirin daily for eight to ten years decreased death from colorectal cancer by up to 35 percent. The study reported an overall 21 percent decrease in death by cancers of all forms noted when taken for more than five years.[59] This result is but one conclusion, but there could be a promising connection between aspirin and the prevention of colorectal cancer.[60]

A controversial origin story, an overlooked inventor, international espionage, pain relief, and heart attack prevention? Not bad for a molecule humankind has known about, albeit indirectly as a precursor, for thousands of years thanks to the bark of the willow tree. Now let's turn our eyes to our next pharmaceutical, lithium, as we uncover how we came to use this puzzling metal in psychiatric medicine.

CHAPTER 4

Lithium

The structure of lithium citrate, one of a variety of chemicals used to give a patient lithium. *Structure generated by author in ChemDoodle v 11.5.0, using information supplied by the National Institutes of Health*

Lithium is different from the other drugs we have covered so far. Lithium is a part of pop culture. Lithium is stigmatized. Lithium is also the title of Nirvana's third-best song. But we almost did not have lithium to rely on for those in our population who suffer from very real mental illnesses. We do, however, thanks to the efforts and guesswork of a very resourceful physician practicing in relative obscurity in Australia, and he did it all by making use of literal guinea pigs in his research that took place in part in the backyard behind his house.

The Discovery of Lithium

Lithium is a little bit different than the small-molecule drugs we have looked at thus far, as it is the element lithium itself that is responsible for the improved medical status of patients who take the medication. Giving patients elemental lithium is frowned on, so treatment of patients with lithium is usually dosed in two different salt forms—lithium carbonate, which is available in an instantaneous-release tablet and extended-release tablet form, and lithium citrate, which is a liquid. Both lithium carbonate and lithium citrate are salts that readily dissolve in the body, freeing the lithium ions to go about through the body and do their job. Lithium is still a small molecule—an ion of lithium is about seven daltons, with its two most popular medication forms, lithium carbonate and lithium citrate, coming in at just under seventy-four and 210 daltons, respectively. Lithium is a metal as well, making for another point of difference with the drugs we have looked at so far, which consisted of just five elements—carbon, nitrogen, sulfur, oxygen, and hydrogen—connected in different arrangements.

The element lithium was not discovered until 1817, with the element's name derived from the Greek word for "stone," λίθος, a word transliterated as "lithos."[1] Nearly a century and a half after its discovery would pass until Australian physician John F. J. Cade changed the face of psychiatry through a series of experiments with literal guinea pigs and patients at the psychiatric hospital where he practiced. Cade trod in the footsteps of his father, David Cade, a physician and World War I veteran who suffered bouts of "war weariness" on returning home. His mental health and the Spanish flu's effect on his family practice left the elder Cade seeking a release from recurring cycles of exhaustion, a release David Cade found in selling his medical practice and taking a position in the mental hygiene department of a local hospital. Against this backdrop, the younger Cade would earn high honors and finish medical school at Melbourne University by the age of twenty-one. During the course of his studies—and akin to Alexander Fleming—John Cade used an inheritance he received in 1931 from a recently deceased uncle, also a physician, to repay his father for the cost of his schooling. John's age put him in the prime position to be called to serve in World War II. In between medical school and the war, Cade worked at the Royal Children's Hospital in Melbourne, Australia, and suffered a bout of bilateral pneumococcal pneumonia that nearly killed him. He lived, thankfully, with the pneumonia fostering the setting wherein Cade would meet his future wife, Jean Charles, who worked as one of the nurses who cared for him as he regained his health. Cade left Australia at the age of twenty-nine in 1941 to serve in World War II, with three years of his service spent as a Japanese prisoner of war in Singapore's Changi Camp.[2]

Of Guinea Pigs and, Well, Guinea Pigs

While interned at Changi, Cade worked tirelessly among his fellow soldiers as a physician. He took note that all his psychiatric patients who died had some level of pathology on autopsy. This observation led Cade to believe there was a toxic agent behind their plight, particularly in those with manic-depressive illnesses. Cade believed the body was making excess amounts of a chemical agent normally found in the body to the point of intoxication in patients with mania and making not enough of a chemical agent when patients presented with melancholia—a set of symptoms we would call clinical depression today.[3] Bipolar disorder was called manic depression in previous generations, and when talking about bipolar disorder in the context of its early research, I will use the terms "manic" and "manic depression" throughout not to be insensitive but to use the language of the time and paint a picture of the stigma revolving around the disorder (a stigma that sadly persists to this day). Although multiple forms of bipolar exist—including a rapid-cycling variant—most forms share in common a fluctuation in "poles" of mood—moments that are very high and euphoric in nature and deep lows with very real consequences without an outside stimulus necessarily needed to cause one of these periodic highs or lows.

Cade returned from the prisoner-of-war camp and went back to private life in Australia. Back home, Cade worked his way up to the position of senior medical officer in the Victorian Department of Mental Hygiene at a war veterans repatriation mental hospital in Bundoora, a suburb of Victoria.[4] Both the hospital and Cade specialized in chronic mental illnesses, with Cade turning his eye again to search for the existence of a toxic agent that led to the condition of his patients. Despite the lack of a bona fide scientific research laboratory, Cade began experimenting with guinea pigs in 1947 to see if there was something in the urine of manic patients that caused their bouts of mania. Cade's idea? Inject the urine of the patients into the guinea pigs and see if the animals started behaving oddly. Urea and uric acid are solid wastes excreted by the body in urine, and Cade believed the secret to mania lie in these wastes.

Along with manic patients, Cade injected urine from patients suffering from melancholia and schizophrenia into guinea pigs, as well as urine from patients without such maladies, as a control. Cade's lab was a woodshed that sat on the hospital grounds or in an abandoned kitchen, depending on the account.

Regardless, both are pictures far from a world-class setting for research. Cade, his wife, and their four children also lived on the hospital grounds, with stories circulating of Cade storing the urine of manic patients in his family's refrigerator and keeping the guinea pigs used for his experiments in a garden in their backyard.[5]

Cade injected urine into the guinea pigs intraperitoneally, that is, into the abdomen, with Cade watching the reactions of the guinea pigs in the hours that followed from a wooden shed.[6] As it turned out, the urine of manic patients was especially toxic to the guinea pigs, killing the animals at lower volumes than the urine from control patients and those with melancholia. The manner of the guinea pigs' deaths was quite terrifying: tremors followed by loss of coordination, leading to paralysis and then seizures before finally succumbing. To further investigate this unusual result regarding the manic patients' urine and in hopes of determining the exact toxic agent responsible for the deaths of the guinea pigs, Cade injected the guinea pigs with solutions of urea, uric acid, and creatine, the three primary nitrogenous solids found in urine. The guinea pigs injected with urea died just as those injected with the urine of manic patients, while those subjected to injections with urea and creatine at the same time did not suffer seizures, suggesting the creatine had a protective effect against urea.[7]

Serendipity stepped in as Cade chose a liquid to deliver the uric acid to the guinea pigs. Uric acid, a solid, is notoriously hard to dissolve in water, but Cade used his keen mind and chose the most soluble of the urates (a name for the salt forms of uric acid), lithium urate, in which to dissolve a sample of urea.[8] Lithium urate is very similar in makeup to uric acid; it features the addition of a lithium atom coordinated to an oxygen atom. When Cade injected the guinea pigs with an 8 percent urea solution also containing lithium urate, he observed a dramatic decrease in toxicity. If lithium urate is never chosen for these experiments—let's say Cade chooses potassium urate instead—the outcome (and the world as we will see) is markedly different.

Thinking the lithium ion to be the reason for the protection against the urea in the injections, Cade tried the experiment again using lithium carbonate and an 8 percent urea solution, with all ten guinea pigs injected with the solution surviving. In contrast, guinea pigs injected with an 8 percent urea solution sans lithium carbonate died in five out of ten cases. When Cade injected the guinea pigs with only lithium carbonate, he observed that the guinea pigs became unresponsive and lethargic for one to two hours before going back to their relatively normal rodent life.[9] Cade certainly had evidence that lithium was serving to protect the guinea pigs from the presence of urea as well as sedate them, but where could he go from here? The results of the guinea pig trials screamed for human participants to test the effectiveness of lithium on, but Cade needed to prove the lithium salts to be safe for humans first.

Before injecting human patients with lithium salts, Cade used himself as a test subject. He ingested the lithium salts at test dosages he planned to later make use of and found no issues.[10] This is a remarkable feat in and of itself with what we now know about lithium dosing, and things could well have gone awry here to the serious detriment of Cade's health. Satisfied with the safety of the prospective dosages, Cade set out to treat a group of nineteen patients in 1948 and 1949—ten patients with mania, six with schizophrenia, and three suffering from melancholia—with lithium citrate. It is important to note that Cade's human experiments lacked a control group, just as Lawrence Craven's aspirin experiments lacked a group not taking aspirin to truly gauge the drug's usefulness.

Cade's chosen test subjects ranged in age from forty to sixty-three, and among the manic patients were those with the most pervasive and disruptive symptoms in the hospital. Cade's pride and joy in the paper is introduced as "W. B.," a fifty-one-year-old male who lived in a chronic state of mania for the past five years. W. B.'s state was so extreme that he was separated from the rest of the population of the hospital. Cade began treating W. B. with lithium citrate on March 29, 1948. Within three weeks, W. B. was allowed within the convalescent ward of the hospital, and on July 9, 1948, he was discharged from the hospital on a maintenance dose of lithium carbonate. To top it all off, W. B. was soon back at his former job. W. B. continued on the maintenance dose until he willfully decided to cease his regimen just before Christmas 1948, leading him back to Cade on January 30, 1949. Once back on a lithium schedule, W. B. was able to return to home and work on February 28, 1949. Cade also tells a few anecdotes in the case of "W. S.," a "powerfully built man of forty-seven years" with recurring manic phases since the age of twenty-five. Due to his size, Cade was grateful W. S. had a pleasant disposition by nature, noting W. S. never once became violent during the height of his illness. This gentleman recovered within a month of beginning treatment with lithium citrate, with a longtime friend saying that "he has never seen him as normal" as when W. S. was under Cade's treatment.[11] While all manic patients in the study showed improvement, the six schizophrenic patients and three patients with melancholia did not show significant signs of recovery.

In the course of the study, Cade noted that the solid lithium carbonate was easier on the gastrointestinal tract than lithium citrate, allowing him to change the administration method of lithium when side effects arose.[12] It is of importance to note that Cade carried out this research without the ability to determine the blood serum concentration of lithium in his patients, a necessity with lithium administration today. Today, patients prescribed lithium salts are closely monitored, with blood samples taken to accurately gauge the concentration of lithium in their blood. Due to the narrow therapeutic window of lithium, a small increase in concentration can wreak devastating effects on the body and lead to acute lithium intoxication, symptoms of which include nausea, vomiting, tremors, and twitching. In severe cases of lithium intoxication, hemodialysis is necessary to remove the excess lithium from the patient's bloodstream. Lithium is concentrated in the thyroid and excreted by the kidneys, with incorrect dosages or decreased kidney function leading to lithium poisoning. Because of this, kidney and thyroid function are routinely checked in lithium administration, as lithium can lead to detrimental effects on both, including nephrogenic diabetes insipidus, renal failure, and hypothyroidism.[13]

Lithium Poisoning and the Slow Journey to Acceptance

Cade's evidence, just as Lawrence Craven's observations on aspirin preventing heart attacks, would take decades for the scientific community to give the merit it deserved. This is not all Cade's fault, as the paper "Lithium Salts in the Treatment of Psychotic Excitement" was published in the September 3, 1949, edition of the obscure *Medical Journal of Australia* within months of a scare concerning the use of lithium in the United States. The culprit? Westal, a table salt substitute consisting of 25 percent lithium chloride.[14]

Using lithium chloride as a substitute for table salt—sodium chloride—has a degree of scientific rationale to it. Lithium is just above sodium on the periodic table, and therefore lithium should have similar properties to sodium. In addition, lithium readily forms a salt with chloride ions, so, from a chemical point of view, using lithium chloride makes sense, especially if it tastes like run-of-the-mill table salt. From a physiological perspective, however, lithium chloride is far from a successful swap, and, in fact, it is quite deadly when used at the quantity necessary to season food. The resulting tragedies of lithium chloride use in the United States likely tossed away any chance of mainstream acceptance that John Cade's obscure but now seminal paper would have in 1949. Why? In response to the very publicized Westal incidents, the U.S. Food and Drug Administration (FDA) recalled all table salt substitutes containing lithium in the same year.[15] Lithium chloride was far from the only consumer product containing lithium in this era. The soft drink 7-Up started out as "Bib-Label Lithiated Lemon-Lime Soda" and included lithium citrate as one of its seven original ingredients, with the "up" in the name intended to be the boost one felt from the lithium. 7-Up continued to give its consumers a boost until 1948, when lithium citrate was removed from the ingredients due to an FDA ban on the use of lithium as an additive in beer and soft drinks.[16]

Like Alexander Fleming and penicillin two decades before him and 10,000 miles away, John Cade was unable to further his lithium research after his

1949 paper, although he did investigate the use of other metal salts to produce a sedative effect like the one seen with lithium carbonate and lithium citrate. Ever humble, Cade made mention of these investigations in passing at a 1970 symposium held in his honor. Among the metals he researched were rubidium and cesium, both natural choices (as they are in the same column as lithium on the periodic table), as well as strontium and the rare earth elements cerium, lanthanum, neodymium, and praseodymium. Only strontium carbonate showed any success.[17]

The lithium salts Cade worked with, lithium carbonate and lithium citrate, were well-known chemical entities and discovered decades before his application to manic patients. As such, many were no longer economically viable, leaving academic research sources to carry out trials to determine their efficacy, as there was little financial incentive for pharmaceutical companies to do so. Without the work of Danish psychiatrists Poul Christian Baastrup and Mogens Schou, who performed similar studies to Cade but with a control group, any chance of using lithium salts in the treatment of what we now refer to as the set of symptoms Cade was observing—bipolar disorder—may have been lost.[18] In a 1982 lecture honoring John Cade, Schou made a comparison of Cade to Baastrup and himself, declaring it was the difference between two different types of scientists: an artistic scientist (Cade) and a systematic form of scientist (Baastrup and Schou). Schou posited that the artistic scientist combined acute observation, inquisitiveness, a desire to test improbable hypotheses, and fearlessness in the eyes of making a fool of himself, while the systematic scientist stood at the ready to confirm the artistic scientist's findings.[19] Medical history and future patients are glad to have both. Thanks to the contributions of Schou, Baastrup, Cade, and others, the FDA approved lithium in 1970 for the treatment of acute mania, with its approval for the prevention of recurring mania in bipolar disorder coming four years later.[20]

Although John Cade is why we have a scientific reason for the use of lithium for bipolar disorder, the use of lithium as a therapeutic in the fight against mental illness did not start with Cade, as a handful of physicians who practiced long before him made use of the element with varying degrees of success. Just before the midway point of the nineteenth century, London-based physician Alfred Garrod repeatedly noticed the blood of patients with gout contained uric acid.[21] Lithium solutions dissolve uric acid crystals—also called urate stones—in a laboratory setting, so it made sense at the time for Garrod to use lithium solutions to treat the uric acid crystals so common in gout.[22] Unfortunately, the dose necessary to dissolve the stones is toxic to a person, leading to a number of side effects. Garrod did not know this at the time, however, and persisted in giving patients solutions of lithium for what he termed "gouty mania" and "brain gout"—what we would now call bouts of depression or manic episodes—as treatment for a working theory he called uric acid diathesis.[23] Garrod's uric acid diathesis suggests a buildup of excess uric acid in the body, particularly the brain, could be the source of the manic or depressed states patients experienced.[24] Decades later, Civil War–era U.S. Surgeon General William Alexander Hammond notes the use of lithium-containing compounds in his 1881 book *A Treatise on Diseases of the Nervous System*. In the book, Hammond makes four references to the use of lithium in its bromide salt form (as well as potassium bromide and sodium bromide), but the sedative effect achieved could just have easily been from the bromide ion instead of the lithium ion.[25] It is of note that there is evidence that lithium was given to "overly excited" soldiers in the American Civil War, but we do not know the form used.[26] An explicit reference to lithium carbonate is made in the writings of Danish psychiatrist Frederick Lange, who used the solid lithium salt in 1894 to treat the melancholia of thirty-five patients in his practice.[27] Little is made of the use of lithium after Lange, possibly due to the toxic effects of lithium if taken in too high a dose and the lack of sufficient technology to monitor the concentration of lithium in a patient's body at the time. Additionally, a plethora of mineral springs contain naturally occurring amounts of lithium, including a set of mineral springs in Mineral Waters, Texas. These springs were unfortunately called "crazy waters" by locals as they witnessed symptoms subside when certain individuals drank from the springs.[28]

In the seventy-plus years since Cade's pivotal study, scientists have yet to determine the exact mechanism of action by which lithium works in patients to stabilize mood. Several proposed mechanisms exist, with these easily filling a few chapters of a book, and it may be that multiple mechanisms come together to bestow lithium's saving grace on those responsive to it for bipolar disorder. One of the broadest possible explanations of how lithium behaves in the body comes through its similarity to magnesium. The lithium ion, the form of lithium available in the body after a lithium-containing medication is taken in and broken down, has approximately the same ionic radius as that of a magnesium ion (0.76 to 0.72 angstroms).[29] Magnesium ions are prevalent in the body and are the backbone of several cellular processes. The similarity in size could allow lithium ions to "trick" the body in situations where cellular mechanisms rely on a magnesium ion, leading to a number of different effects on the body, so many that it is hard to tell which ones are responsible for the bipolar disorder–fighting properties of lithium.[30] A big difference between the lithium and the magnesium ion persists, however, as a magnesium ion is doubly charged, while a lithium ion has a single charge. Animal studies show lithium to increase the amount of serotonin—a neurotransmitter that regulates mood and feelings of reward—synthesized by the body as well as its subsequent release.[31] Lithium may also have a neuroprotective effect on the brains of patients, with imaging studies of patients treated with lithium showing an increase in the size of the cerebral cortex and hippocampus.[32] How these effects happen is still unknown, but it likely hinges on the protein complex Akt/GSK-3β, which is altered in the presence of lithium ions, with the protein Akt then able to act on additional proteins that promote membrane plasticity, cell cycle progression, and motility.[33]

One of the major successes of lithium as a therapeutic agent is the very real decrease in risk of suicide for patients with bipolar disorder or other mood disorders when taking the drug. Those living with bipolar disorder commit suicide at a rate as high as 10 percent, ten times the suicide rate in the general population. This 10 percent chance rises steeply to 25 percent for male bipolar patients with a history of self-harm.[34] With lithium therapy, the suicide rate drops to a little less than double the rate of the general population.[35] In addition to the decline in the suicide rate, per a 1994 study, it is estimated that lithium therapy has saved the United States an estimated $145 billion for the long-term hospitalization of manic-depressive patients, with this number adjusted to more than $250 billion in today's dollars.[36] Lithium, despite the necessary blood monitoring and ever-looming threat of toxicity, helps many people go back to at least some version of their normal lives (as we could see especially well early on with W.B. in John Cade's 1949 study), and that is thanks to Cade's serendipitous decision based on the solubility of uric acid in lithium urate coupled with a keen eye for observation and a care for the underlying cause of his patients' infirmities.

CHAPTER 5

Iproniazid

The structure of iproniazid. *Struc-ture generated by author in Chem-Doodle v 11.5.0, using information supplied by the National Institutes of Health*

Unlike penicillin or some of the other drug therapies in this book, you have likely never heard of iproniazid. This medication started life as an agent in the fight against tuberculosis, only later to be discovered to have an interesting side effect. Patients became giddy shortly after taking iproniazid. This side effect would soon become iproniazid's primary use and lead it to become the first in a class of antidepressants called monoamine oxidase inhibitors (MAOIs).

From Isoniazid to Iproniazid

It took a couple of benchtop chemistry steps to get to the chemical compound iproniazid from its original precursor. In 1951, two different research groups, one entrenched in the academic world at Squibb Institute in New Jersey and the other in the pharmaceutical conglomerate Hoffman-La Roche, discovered the effectiveness of isonicotinyl-hydrazine against tuberculosis in animal models within months of each other. Part of this is because they likely started with a similar synthetic precursor, hydrazine. Why were they working with versions of hydrazine? Large amounts of hydrazine left over from World War II, wherein it was used by Germany as an alternative to ethanol to fuel V-2 rockets. The stockpiles of hydrazine were discovered and offered to pharmaceutical and chemical companies at little cost to use as precursors in organic synthesis reactions.[1] Despite their success in applying it to solve the problem of tuberculosis, neither the Hoffman-La Roche nor the Squibb Institute team was the first to synthesize isonicotinyl-hydrazine. The molecule was first synthesized by Hans Meyer and Josef Malley at the German Charles-Ferdinand University en route to finishing a PhD thesis.[2] Both the Squibb Institute and Hoffman-La Roche initiated clinical trials that were successful in showing isonicotinyl-hydrazine could treat tuberculosis in humans. Tuberculosis is a highly infectious disease, most commonly centered on the lungs and caused by the bacterium *Mycobacterium tuberculosis*. Patients with tuberculosis suffer through weeks of bloody coughing fits, chest pain, fever, night sweats, and drastic weight loss. A little less than a century and a half ago, tuberculosis was responsible for one in every seven deaths in the United States and Europe. If we go back just a little farther and look at seventeenth- and eighteenth-century Europe, tuberculosis led to one in every four deaths.[3] Tuberculosis ravaged humanity for millennia, with our earliest evidence of its infection in humans coming in traces of *M. tuberculosis* DNA found in skeletal remains dating to roughly 7250 BCE. The DNA was found in the rib and arm bones of a mother and the long bones of her infant child buried together at a submerged site, Atlit-Yam, in the Mediterranean Sea.[4] Only in the past century have we gained a foothold against this killer, successfully enough so that a tuberculosis skin test to see if you show the telltale "bump" of being exposed to *M. tuberculosis* has become a routine step before going off to college or starting a new job in the health care industry.

In the 1950s, streptomycin existed to combat *M. tuberculosis*, but resistance to the antibiotic was a problem.[5] Lung collapse therapy—the intentional injection of excess nitrogen into the lungs—was also an option, but whether the procedure fulfilled the hypothesis that it gave the lungs time to rest and heal was perpetually in question.[6] Prior to these two therapies, a trip to a sanitarium for fresh air and rest in an outdoorsy setting was the preferred remedy but was available only to the well to do. Sanitariums were wildly popular for those who could afford the several months' stay and time away from work, leading to an explosion of sorts in the field in the first quarter of the twentieth century. During this time, the United States saw its total number of sanitarium beds increase from approximately 5,000 in 1900 to nearly 675,000 by 1925.[7] As a sanitarium stay was not an option for everyone, a better answer to the tuberculosis problem was needed, and clinical trials showed that isonicotinyl-hydrazine could be the answer. After the trials concluded, isonicotinyl-hydrazine soon earned the name of isoniazid. Not long thereafter, however, it was usurped by the application of a common organic chemistry maneuver—the addition of an isopropyl group (three carbons with their accompanying hydrogens)—to isonicotinyl-hydrazine by scientists at Hoffman-La Roche, yielding a more effective drug that they named iproniazid.[8] The molecular mass of isoniazid is 137 daltons, while the molecular mass of iproniazid is 179 daltons, both well within the constraints of modern small-molecule drug design. If Hoffman-La Roche leveraged their discoveries correctly, the company would have a hit on their hands in iproniazid. Sold under the brand name Marsilid, iproniazid gave physicians another much-needed armament in the fight against tuberculosis.

Iproniazid's New Role

Hoffman-La Roche, however, did not suspect the next chapter in iproniazid's life. Not long after psychiatrists began prescribing iproniazid to patients for their tuberculosis, Irving Selikoff and Edward Robitzek of Sea View Hospital started to see an interesting side effect—an increase in pep and even euphoria in patients.[9] A now-classic Associated Press picture taken of Sea View Hospital tuberculosis patients from this period shows a group of sanitarium patients dancing, with the foreboding description, "A few months ago, only the sound of TB victims coughing their lives away could be heard here."[10] In the same year, Ernst Zeller's research group at Northwestern University discovered that iproniazid would irreversibly bind and inhibit monoamine oxidase, making iproniazid the first in a long line of MAOIs. Why is this important? Monoamine oxidase is an enzyme that breaks apart the neurotransmitters serotonin, dopamine, and norepinephrine. Inhibiting monoamine oxidase allows for an increase of serotonin and dopamine in the brain, leading to an antidepressant effect. This effect can last for weeks after the drug is no longer being taken, as the body has to create all new molecules of monoamine oxidase since the existing ones are irreversibly bound in the presence of iproniazid.[11] Zeller also noted that isoniazid, iproniazid's precursor, did not bind monoamine oxidase.[12] In time, this euphoric "side effect" of iproniazid would be of more interest than its intended use in fighting tuberculosis, as several studies were conducted to monitor the mood-altering effect of iproniazid in tuberculosis patients. One of the researchers in these studies, Max Lurie, likely used the term "antidepressant" for the first time to refer to iproniazid's unanticipated action.[13]

A Company Questions the Future Size of the Antidepressant Market

Some research was missing, however, in that no explicit data existed that showed iproniazid's effects on patients without tuberculosis. Nathan Kline, a psychiatrist at Rockland State Hospital in Orangeburg, New York, led a research team in a study of twenty-four individuals: seventeen with schizophrenia and seven with depression. The patients were given fifty milligrams of iproniazid three times a day, with seventeen of the patients showing significant improvement after taking the drug. Kline, ecstatic about the success and possibilities of the future application of iproniazid, deemed the drug a "psychic energizer" as he explained the drug's action.[14] This extra energy was big in and of itself, as most psychiatric drugs at the time sought only to sedate the patient.[15] Hoffman-La Roche, which still held control of iproniazid due to its success as an antituberculosis drug under the brand name Marsilid, almost cut ties with Kline and his research in 1957 as several of the higher-ups within the company believed the antidepressant market to be too small and unpredictable to warrant any effort. Hindsight shows this to be an almost comical supposition. Kline and his research group from Rockland State Hospital saved their ongoing research in an eleventh-hour plea to L. David Barney, president of Hoffman-La Roche, over dinner. Unfortunately, squabbles over credit for the discovery of iproniazid's antidepressant properties tore Kline's group apart later in the year.[16]

Hoffman-La Roche never sought approval of Marsilid for depression. Despite this, off-label use of Marsilid continued, with more than 400,000 individuals treated for depression with the tuberculosis drug and first MAOI inhibitor.[17] This was a boon to the psychiatric community, as the most popular treatment at the time was electroconvulsive therapy (ECT), the passage of electric current through the brain of a patient while they are under general anesthesia.[18] The electric current causes seizures in the brain, which, in theory, alters the chemistry of the brain in hopes of eliciting a change in behavior or feeling. This idea of changing brain chemistry is not all hocus-pocus, as they knew at the time ECT increased the concentration of serotonin in the brain, at least in animal studies.[19] Needless to say, having an antidepressant in tablet form was greatly appreciated by those who might otherwise be facing an ECT session.

The End of Iproniazid

In 1961, iproniazid was pulled from the market in the United States due to the prevalence of occurrences of hepatitis in patients taking the medication.[20] Iproniazid forms free radicals when metabolized, with these free radicals possibly leading to the liver damage incurred while taking the medication.[21] Iproniazid would remain on the market in Canada until 1964, when it was removed for a peculiar set of interactions known as the "cheese effect," which occurs when someone taking iproniazid or another MAOI eats or drinks an item with a high amount of tyramine in it, like cheese, wine, chocolate, a variety of caffeinated drinks, or smoked meats. Tyramine spurs on the body to release norepinephrine, which is normally broken down by monoamine oxidase. In the presence of an MAOI like iproniazid, however, this release of norepinephrine is not regulated by the body's system of checks and balances and broken down, leading to terrible headaches and rapidly increasing blood pressure in a patient.[22] Tyramine is prevalent in a plethora of foods, making the cheese effect difficult for patients to avoid. Iproniazid was eventually replaced by two different MAOIs, phenelzine (brand name Nardil) and tranylcypromine (brand name Parnate), with the pair accounting for 90 percent of the market well into the 1980s.[23] MAOIs as a class have fallen out of favor due to their high number of side effects, particularly interactions with tyramine and the subsequent cheese effect, and the arrival of selective serotonin reuptake inhibitors as a class of drugs to treat depression.[24] Interestingly, iproniazid's precursor, isoniazid, is still used to treat tuberculosis, which still sickens more than 10 million people a year and leads to nearly 1.5 million deaths annually.[25]

CHAPTER 6

Digoxin

The structure of the quite complex digoxin. *Structure generated by author in ChemDoodle v 11.5.0, using information supplied by the Royal Society of Chemistry*

Digoxin is quite the enigma. The drug holds both the power to heal and the power to harm, as we will see explicitly through the actions of former nurse Charles Cullen later in this chapter. In the right dose, digoxin increases the efficiency of the heart, but too much and it can send a patient to death's door. Digoxin is an extract of the *Digitalis lanata* species of the foxglove plant. Human beings knew the foxglove hid a medical secret since the 1500s, but safely revealing this secret took the work of several masters and the course of hundreds of years.

Of Foxglove and Its Champion, William Withering

Foxglove itself has a mysterious nature to it, with the name "foxglove" a reference to either fairies or a long-forgotten musical instrument that featured bells arrayed on an arched structure. The origin behind the scientific name of foxglove is better known, dating to the sixteenth century and a Bavarian physician and hobby botanist by the name of Leonhart Fuchs, who, noting the finger-like flowers of the plant, lent it the genus name *Digitalis*. Fuchs also made use of foxglove in his practice per his 1542 book *De Historia Stirpium Commentarii*, albeit for the reaction elicited when dosed to the point of extremes, where Fuchs describes foxglove as a purgative and emetic that might be useful in the treatment of dropsy. Dropsy is an older term for what we now call edema. Edema is a swelling of the body, particularly due to the heart pumping weaker than it should, leading to liquid pooling in a person's extremities. *Digitalis* appears in the 1661 *London Pharmacopoeia* but was often overlooked during the period due to its toxic side effects, which rear their head with slight changes in dose.[1]

Foxglove would not be overlooked for long, however, thanks to the tremendous efforts of a renaissance man by the name of William Withering. Withering, a physician by trade, excelled in a number of fields, including botany and minerology (Withering even has a mineral prevalent in northern England named after him, Witherite, a combination of barium, carbon, and a trio of oxygen atoms). Withering's explorative mind was balanced by a love and concern for his fellow man—he helped establish a clinic for the poor in Staffordshire, England, before the bigger city of Birmingham came calling and he took a position in the city's General Hospital. This move came at the behest of his mentor, the acclaimed physician Erasmus Darwin, grandfather of naturalist Charles Darwin. Withering would become acquainted with the foxglove plant by an unusual coincidence. In 1775, a friend brought Withering a recipe for a medicine purporting to cure dropsy consisting of more than twenty herbal ingredients. Calling on Withering's background in botany, he asked the doctor if he knew which one ingredient was the key to the cure. Withering immediately singled out foxglove as the active component in the mixture. This chance encounter begins a ten-year journey into the application of foxglove and, in particular, the species *Digitalis purpurea* in humans, with Withering carefully documenting its administration to 163 patients and publishing his findings as *An Account of the Foxglove, and Some of its Medical Uses: with Practical Remarks on Dropsy, and Other Diseases*.[2]

One of Withering's key observations came in noticing foxglove was useful to patients who developed dropsy after a bout of scarlet fever or a set of symptoms we now associate with strep throat. Both are caused by bacteria of the *Streptococcus* genus, and we now also know that severe infections of this nature can lead to heart damage, particularly to the valves, which would undermine the heart's ability to pump efficiently and lead to swelling.[3] Withering described foxglove as having "a power over the motion of the heart to a degree yet unobserved in any other medicine," easing the irregular and rapid heartbeats seen in atrial fibrillation and atrial flutter.[4]

Withering did not go around prescribing foxglove for everything but used it as a last line of defense in treatment when nothing else would work. In his book, Withering noted the toxic effects of foxglove, wherein it leads to vomiting, eye issues (including green to yellow vison), and an extremely slow pulse. He recorded a pulse as low as thirty-five beats per minute in one patient.[5] He paid particularly close attention to dosages and the condition of the foxglove leaves he gathered, as he knew this was vital to potency. In his writings, he gives a harrowing account of a woman who was aware of the usefulness of foxglove but not the dire situations that arise when too much is given. The woman added a handful of foxglove leaves to a half-pint of water and boiled the mixture, giving her husband what amounted to a foxglove tea in hopes of treating his asthma. Her husband received something entirely different: a nearly lethal dose of foxglove. Withering writes, "This good woman knew the medicine of her country, but not the dose of it, for the husband narrowly escaped with his life."[6] Withering initially believed these side effects seen with foxglove, particularly vomiting, to be a sign the foxglove treatment was working, pushing the therapy on the patient to the point and past that of side effects, before revising this point of view later in life and ceasing the dose when the first signs of side effects show.[7]

Stories like the one involving a caring wife and the unfortunately strong foxglove tea led Withering to take steps to standardize the dose of foxglove he gave patients to avoid intoxication. To do so, he gathered foxglove leaves at only one time of the year and only when the plant was flowering.[8] Not only do the leaves of the foxglove contain the active ingredient, but so do the flowers, seeds, and sap, just to a lesser extent.[9] Withering then dried the leaves by sunlight or fireside before crushing them into a fine green powder.[10] He then gave the powder to the patient or made an infusion with the powder.[11]

Withering died of tuberculosis at the age of fifty-eight on the cusp of a new century in 1799.[12] Foxglove and thus *D. purpurea*, despite the warnings of Withering, entered into the medical lexicon of the time and saw its application in a variety of means outside of easing the effects of dropsy. The foxglove plant is a point of controversy in the life of Vincent van Gogh, where the plant is sometimes linked to the yellow hue so common in the master's paintings. Such an effect on one's vision is called xanthopsia, which can result from bilirubin building up in the eye in rare cases of chronic foxglove overdose.[13] Van Gogh suffered from a variety of mental and physical ailments and on entering an asylum in 1889 at the age of thirty-six likely received foxglove in some form at the behest of Dr. Paul-Ferdinand Gachet. Why can we be so sure? Foxglove was a sort of cure-all drug at the time, prescribed for anything from mental illness to headaches to eye pain.[14] In addition, one of van Gogh's paintings holds a clue. *Portrait of Dr. Gachet* (1890) posits the physician leaning on a table with a stack of books, one hand propping up his head and the other resting next to a vase holding a sprig of foxglove as Gachet looks off in the distance. While the connection is present, whether foxglove was responsible for van Gogh's use of yellow is disputed. Dr. Gachet knew quite well the dangers inherent in administering foxglove, with the doctor even writing a paper on how to properly dose the medication.[15] Also, van Gogh was under Gachet's care for only two months, not nearly long enough to develop xanthopsia.[16] The paintings of van Gogh prior to 1889, particularly 1888's *Still Life: Vase with Fifteen Sunflowers*, show a disposition to the color yellow, which predates his entering the asylum and the care of Dr. Gachet.[17] While it is possible that van Gogh received an eyesight-damaging amount of foxglove prior to meeting Dr. Gachet, the nail in the proverbial coffin comes in the form of an eye examination performed on van Gogh by Dr. Gachet. The exam showed van Gogh to exhibit nearly perfect vision with no abnormalities in distinguishing one color from another.[18] If Gachet performed this test correctly, any signs of xanthopsia should have been made apparent at the time. It is likely the often-tormented van Gogh leaned on yellow hues in many of his works simply because he favored the color.

Isolating Digoxin from *Digitalis*

Administration of digoxin is commonly called digitalis therapy, with an extract of plants from foxglove and thus the *Digitalis* genus given to make the heart work more efficiently, and the name "digitalis" continues to be used even though we now know and give the active ingredient of the plant that amplifies the action of the heart, digoxin. Digoxin was first isolated in 1930 by Sydney Smith, who sought an answer as to why the leaves of *D. lanata*, also known by the amusing name woolly foxglove, were more physiologically potent in digitalis therapy than the leaves of its cousin, *D. purpurea*.[19] The chemical digitoxin was known to be responsible for at least part of the action of *D. purpurea*, as it was extracted six decades earlier in 1875 by Oswald Schmiedeberg.[20] Schmiedeberg, a renowned German pharmacologist and teacher, received a small amount of fame for testifying in 1911's bizarre case of *United States v. Forty Barrels and Twenty Kegs of Coca-Cola*, wherein the federal government sued a group of inanimate objects to force the Coca-Cola Company to take caffeine—which the company increased in the formula as cocaine was removed—out of its product. The initial case ended on a technicality, with the Coca-Cola Company arguing caffeine was a naturally occurring ingredient and not an additive but not before a 1916 reversal saw caffeine declared an additive by the U.S. Supreme Court. Eventually, the Coca-Cola Company assuaged, agreeing to pay court costs and dropping the caffeine content of its famed drink by half.[21]

Sydney Smith knew of Schmiedeberg's discovery of digitoxin and knew the compound to be present in *D. lanata*, but he did not know why this species of the plant was so potent when used in digitalis therapy. To solve this mystery at hand, Smith sought to separate the digitoxin from the plant to determine if it was a matter of increased concentration or if there was also a similar chemical compound present responsible for the added physiological impact. Thanks to Smith's deft chemistry skills, he was able to separate out digoxin, which was never before isolated and the key as to why the digoxin-containing *D. lanata* is so much more biologically active than *D. purpurea*.[22] Further studies have shown *D. purpurea* to consist primarily of digitoxin, whereas *D. lanata* contains a mixture of both digoxin and digitoxin.[23] Digitoxin and digoxin, as you could suspect, are very similar molecules, with both being members of the cardiac glycoside family of medications.

At the time of this discovery, Smith was employed by the pharmaceutical company Burroughs Wellcome, which is now a part of GlaxoSmithKline.[24] GlaxoSmithKline is still in the business of extracting digoxin, contracting farmers to grow woolly foxglove in the Netherlands before drying the leaves and shipping them to the United States for processing. The company sells the final extract under the brand name Lanoxin.[25] A synthetic method to produce digoxin exists, but it is easier and more profitable to continue to extract the chemical compound from the foxglove plant. I should also mention that digoxin consists of only three different atoms—carbon, hydrogen, and oxygen—but weighs in at nearly 781 daltons, far more than any of our previous pharmaceuticals and the first to cross the 500-molecular-mass threshold outlined by Lipinski's Rule of Five.

Digoxin works by altering the effect of the enzyme sodium-potassium ATPase. This enzyme acts as a sodium pump, moving three ions of sodium out of the cell and two ions of potassium into the cell each time a molecule of adenosine triphosphate (ATP) is hydrolyzed into adenosine diphosphate. Digoxin inhibits the work of sodium-potassium ATPase, leading to an increase in the amount of another ion, calcium, in the heart and thus prodding the heart to work by increasing the rate of contractility of muscle fibers in the heart. This results in an overall increase in cardiac efficiency, restoring to (nearly) normal a heart that has been worked to the point of exhaustion.[26] This is the main use of digoxin as a therapy today. A 2008 study showed digoxin to inhibit HIF-1, a protein that codes for many genes that lead to cancer, but the concentrations of digoxin used in the study were three to ten times higher than the normal plasma concentrations when used in humans. At the high concentrations necessary to elicit a cancer-fighting effect, numerous side effects and toxicities would abound, rendering any positive effect of the larger dose useless.[27] As with many of the medications we have come across, digoxin exhibits a narrow therapeutic window, with miscalculation in dosage leading to serious side effects. The issue of the therapeutic window has made some physicians shy away from its use over the years, but a large-scale study helmed by the Digoxin Investigation Group showed digoxin to reduce hospitalizations in those suffering from congestive heart failure with no negative effect on mortality when dosed properly.[28]

Digoxin as Poison and the
Many Murders of Charles Cullen

Digoxin, along with its cousin digitoxin, was a well-known poison when dosed to the point of extremes. Agatha Christie made use of digitoxin as a method of poisoning in six stories, with Hercule Poirot left to suss out its handiwork in 1937's *Appointment with Death*.[29] Digitalis therapy has played a role in a number of real-life crimes as well, with digitalis playing a central one in an insurance fraud scheme carried out in New York in the 1930s. In the scheme, the patients would take out a life insurance policy with a disability clause. The patients would then be coached on the symptoms of heart failure and administered digitalis by a physician, which led to a recordable change in the patient (and a paper trail) through an abnormal electrocardiogram reading. The patient would then be eligible for disability but owe the physician (and the lawyers who turned the physicians on to the scam) a cut of their windfall. It took insurance companies nearly five years to catch on to the scam, but the ring of physicians and lawyers were eventually prosecuted and convicted by 1941. The use of digoxin for illicit means has done substantially more damage than a little insurance fraud over the years. It was also the weapon of choice of murder of perhaps the most prolific serial killer in history, Charles Cullen.

Charles Cullen was and is a troubled soul. His first suicide attempt came at the age of nine, when he ingested the remnants of a secondhand chemistry set he received from a church donation drive with a glass of milk. Suicide attempts would be strewn throughout his life and become part of an almost constant thought cycle for Cullen. He attempted suicide again during his senior year of high school after the death of his mother in a car accident as she suffered an epileptic seizure. Part of Cullen's depression arose from anger that the hospital would not let him see the body so he could say good-bye to his mother, who shielded him from his seven much older brothers and sisters.[30] Cullen turned down a dark path in hopes of being reunited with his lifelong protector.

Charles escaped, for a time at least. He enlisted in the navy, starting a six-year contract that would see him work as an electronics technician chained to the sixteen Poseidon C3 nuclear missiles carried by the USS *Woodrow Wilson*, a claustrophobia-inducing attack submarine. Cullen's cycle of self-endangerment continued, and one day, one of his superiors, Petty Officer First Class Michael L. Leinen, found Cullen dressed in full operating room scrubs and seated at the controls of the *Woodrow Wilson*'s Poseidon missiles. This would the end of his tenure aboard the vessel, as his mental state did not lend itself to close proximity to nuclear weapons. Luckily, Cullen would not have been able to fire the missiles without additional aid should he have chosen to launch them.[31] After the incident, Charles was sent to a less intense position aboard the USS *Canopus*. Aboard the submarine, Cullen would be spared from spending his days deep under the water. Unfortunately, Cullen's navy contract would continue to weigh on him, propelling him to a total of three suicide attempts during his tenure in the navy along with stays in the Charleston Navy Hospital Psychiatric Ward.[32]

By 1984, Charles Cullen and the navy parted ways. The twenty-four-year-old Cullen was looking for a new beacon to set the course of his life's direction. He eventually found it by attending nursing school—and not just any nursing school but the one attached to the hospital where his mother died. During his time in the navy, Cullen told his bunkmate Marlin Emswiler that he liked helping people, with the possibility of becoming a nurse already rolling around in his mind at the time.[33] He stood out as the only male in the class and was elected to serve as class president. Things were different, as Cullen now appeared to be gaining a sense of traction he had never before experienced. He paid his tuition with shifts at Dunkin' Donuts and Roy Rogers.[34] Charles met a girl, Adrianne Baum, and the two soon married. Charles was in control of his life for the first time.

This control would be the undoing of dozens if not hundreds of lives during Cullen's sixteen-year career as a nurse. During his career, Cullen bounced from nursing position to nursing position, changing facilities nine times, often under suspicious circumstances stemming from the deaths of some of his patients.

Cullen's life trajectory at the start of nursing school did not take long to stall out. In late 1992, Adrianne Baum filed for divorce from Charles Cullen, serving him papers as he worked in the intensive care unit.[35] The grounds? Refusing to sleep with her and cutting off communication with her as well as abuse of the couple's Yorkshire terriers.[36] Cullen, without much money, represented himself in the ensuing litigation and after the divorce moved into a basement apartment. Charles would have likely ended his nursing career at some point in the next few years, but he stayed on, in need of money to pay monthly alimony and child support checks for the two children Baum and Cullen shared.[37]

Shortly after the divorce, Charles Cullen worked at Warren Hospital in Phillipsburg, New Jersey, where Charles became acquainted with digoxin, but in a manner to kill and not to heal. In time, digoxin would become his go-to murder weapon of choice. An unidentified male nurse entered the room of Helen Dean, a ninety-one-year-old recovering from breast cancer surgery. The nurse told her son, Larry, to leave the room, then injected Helen. As one could guess, Helen took a turn for the worse almost immediately, and by the next afternoon, she was gone.[38] However, toxicology scans turned up no signs of any illegitimate medications; of the more than 100 chemicals tested for, digoxin was not included. Cullen and several other nurses were scrutinized and given lie detector tests, but Cullen passed and continued working at Warren Hospital. Cullen would soon move on, bouncing from nursing position to nursing position, free from severe scrutiny until he was in the employ of Liberty Nursing in October 1998. Multiple coworkers saw him enter the room of an elderly female patient carrying syringes, a patient for whom he was not assigned to care. The patient did not expire but did endure a broken arm in what we can only surmise was some sort of altercation.[39] Liberty Nursing immediately fired Cullen, but Cullen continued to be able to find work, partially because of a nationwide nursing shortage and partially because of a lack of detailed information sent between hospitals regarding potential hires. Cullen was hired in less than a week, this time by Easton Hospital in Pennsylvania. By year's end, Cullen would be tied once more to a patient's premature passing. This time, it would be Ottomar Schramm, a seventy-eight-year-old retired steelworker who received a toxic amount of digoxin, a medication he should have not been given because Schramm had a pacemaker. After this incident, Cullen stayed on at Easton for a short time while also working part-time at nearby Lehigh Valley Hospital before transitioning to Lehigh full-time in March 1999.[40]

Charles succeeded in killing because he used relatively benign and under-the-radar pharmaceuticals, like digoxin and insulin, to carry out his murders. No high-dollar or scheduled drugs showed up missing, and no morphine that needed to be carefully monitored showed up missing on a nightly checklist and raised a red flag. Charles was smart in this manner. Digoxin kills by severely altering the rate and rhythm of the heart and its beating, often lowering the heart rate to the point of permanent damage (remember Withering's observation of a thirty-five-beat-per-minute heart rate in a patient).

Charles Cullen's involvement in the murders of numerous patients did not quiet his suicidal thoughts, and on January 2, 2000, at the dawn of a new millennium, Charles entered his bathroom and set up a charcoal grill. He carefully insulated the vents to seal off the room, lit a fire, and tried to asphyxiate himself via carbon monoxide poisoning. Charles was saved from a death in the coffin of his basement apartment when his neighbor Karen Ziemba smelled kerosene and called the police.[41] Grief, turmoil, despair, and death continued to follow Cullen, only the latter was never his own.

As Charles Cullen advanced in his career and hopped from hospital to hospital, he used a knowledge of systems and electronics—no doubt instilled in part while in the navy—to manipulate hospital policy and computers. A perfect example is his deft work around a Pyxis MedStation later in his nursing career. A Pyxis MedStation is a sophisticated dispensing device that houses medication and is connected to a terminal where a nurse, pharmacist, or doctor can input orders to receive a medicine within. Think of it as like a vending machine on steroids but with steroids in it as well. Cullen determined that if he put an order into the Pyxis MedStation for a drug and then immediately canceled the order, the door containing the drug would still open. This placed the drug Cullen ordered right within his grasp and left zero paper trail for others snooping around to follow up on.[42] In time, this glitch in the system would become suspicious in and of itself, with others noticing the plethora of canceled orders Cullen was inputting, but Cullen found another way to get what he needed using the Pyxis machines. By some stroke of luck for Cullen, acetaminophen and digoxin shared a drawer in the Pyxis machine he was using. One could ameliorate a headache; the other could kill. Cullen simply put in an order for acetaminophen but grabbed digoxin as the door opened.[43]

After Lehigh Valley came the cardiac unit at St. Luke's Hospital in Bethlehem, New Jersey, a quizzical locale for one such as Cullen. Why was Cullen so successful in being hired time after time despite the suspicious circumstances often predicating his departures? Hospitals and prospective whistle-blowers likely feared Cullen would claim slander and take legal action. This did not, however, stop one brave cadre of nurses at Sacred Heart Hospital, who offered to tender their resignation should Cullen remain in the employ of the hospital.[44]

At this juncture in Cullen's story, his fellow nurses begin daring to speak out, and the state police of New Jersey begin initiating investigations. Why? Possibly the buildup of rumors amidst Cullen led to a tipping point, but one incident in particular spurred the attention. A fellow nurse found unopened vials of procainamide and sodium nitroprusside stashed in a sharps container (a disposal for used syringes) and alerted superiors.[45] Both procainamide and sodium nitroprusside follow Cullen's modus operandi of using inexpensive drugs and drugs that are not able to be recreationally abused, but both can lead to death in the case of an overdose. Procainamide alters the action of the heart and can lead to a sharp increase in heart rate, while sodium nitroprusside can lead to an abrupt decrease in blood pressure. Sodium nitroprusside, due to its chemical makeup, can also lead to cyanide poisoning if overdosed.[46] An investigation showed that Cullen was likely the one behind the theft and hiding of the drugs, leading him to resign from St. Luke's almost immediately. A fellow nurse called the state police to report the theft and request an investigation, with the records of nearly seventy patients who died on Cullen's watch reviewed, but no connection was found.[47] It was mid-2002, however, and, thankfully, Cullen's time in hospitals was coming to an end. Coworkers took notice of him accessing records for patients he was not assigned to and saw him in the patients' rooms. Cullen found himself in yet another new job in 2003 at Somerset Medical Center in Somerset, New Jersey. That summer at Somerset would be his undoing, however, as the hospital software that benefited him earlier with the Pyxis MedStation mishaps would finally ensnare him. On June 15, 2003, Cullen called up the files of forty-year-old cancer patient Jin Kyung Han. Cullen was not assigned to Han prior to Ms. Han undergoing a cardiac event. Although Ms. Han survived this event, she would be dead within three months. What linked Cullen to the cardiac event was a software trail showing Cullen had ordered digoxin for a different patient just before Ms. Han's episode occurred.[48] Later in the month, individuals saw Cullen in the room of a Florian Gall, a Roman Catholic priest to whom Cullen was not assigned. On June 27, 2003, Cullen called out his former play, ordering digoxin for another patient and then reportedly giving the digoxin to Reverend Gall, who died the next day. Things would play out differently this time. Reverend Gall's sister Lucille suggested foul play on learning her brother's postmortem panels showed signs of digoxin.[49] On Friday, December 12, 2003, Charles Cullen was placed under arrest for one murder and one attempted murder during his tenure as a nurse at Somerset Medical Center.[50] The arrest came in an awkward position for anyone, let alone an early forties divorcé, while Cullen and a romantic interest were enjoying spring rolls and drinks.[51] The next day, Cullen gushed forth to the authorities, supposing he had killed forty patients. Investigators believe the number to be much higher, closer to the 300 to 400 range, making Cullen the most prolific serial killer in U.S. history.[52] Cullen agreed to cooperate with authorities if they would not seek the death penalty. They obliged. This was an oddity in light of the sheer number of people Cullen killed, but prosecutors made the agreement in hopes of finding some sort of closure for the families of his victims through Cullen's cooperation. In total, Cullen pled guilty to twenty-two murders and six attempted murders, resulting in an incarceration term equaling eleven consecutive life sentences. Cullen will have to wait 397 years until he is eligible for parole.[53]

Cullen preyed not only on the elderly but on the young as well, taking the life of a twenty-one-year-old man as Cullen vied for control over something in this world. The twenty-one-year-old was merely in the hospital for a spleen transplant, a medical event typically not considered life threatening.[54] Charles thought he was doing something genuinely good by intervening in and taking the lives of so many. He thought he was stopping a cycle of pain—pain that might not be here yet but would arrive in time. Charles thought, in some perverse manner, he was doing something good, something helpful, something protective.[55]

Despite the murders, a chance did arise for Cullen to do something genuinely good. A letter came to Cullen as he awaited in oblivion for a series of sentencing dates and days in court where he would be put on display for the families of his victims and hopefully show some sign of remorse, a sign that, unlike this fated letter, never came. In the letter was either a chance at redemption or a chance to play God in another individual's life or, from Cullen's point of view, a perverse mixture of the pair. The note within was from the mother of an ex-girlfriend of Cullen's, a girlfriend he shared a child with, and pleaded with Cullen to take the antigen test necessary to see if he was eligible to be a kidney donor for her son, Ernie Peckham.[56] Cullen agreed to test for the one-in-a-million donor match, and he turned out to be a perfect match; receiving a serial killer's kidney would be preferable to a seven-year wait for a donated one. Cullen, energized by the possibility, sought out the donation as if it was his only source of oxygen as he floated aimlessly across the stars. The prosecution and families of his victims simply did not want Cullen darkening the doors of a hospital once more. Why? Because they feared Cullen would use this opportunity to harm another or even commit suicide and pull from their grasp any chance at the justice they wanted on behalf of their loved ones.

Despite the luck of the match, Charles Cullen found a way to endanger Ernie Peckham's life. During a sentencing hearing in Pennsylvania, Cullen became incensed with the judge. Cullen started repeatedly saying, "Your honor, you must step down."[57] He would not stop, not for his well-being, for the crowd filled with the relatives of his victims, or for fear of destroying any chance of donating his kidney. Officers placed a spit mask around his face. Cullen kept yelling. They placed a towel around his mouth. Cullen kept yelling. Finally, they duct taped him, but Cullen continued his ranting. Cullen was already damned by his guilty plea, with the plea tacking on six more life sentences in the courtroom that day. But would his actions be condemning Ernie Peckham to an early death as well if this outburst shot down any chance of the kidney donation? Miraculously, no, thanks in no part to Cullen, other than the hope that lay obscured by the skin, fat, and sinew covering his midsection. In the end, Cullen was spirited away from his cell and successfully donated a kidney to Ernie Peckham in August 2006.[58] As of this writing, it is a decade and a half later—nearly as long as his entire nursing career—and Charles Cullen is sitting in a prison cell somewhere in the United States, a little less than four centuries away from a chance at parole.

CHAPTER 7

Chlordiazepoxide

The structure of chlordiazepoxide. *Structure generated by author in ChemDoodle v 11.5.0, using information supplied by the National Institutes of Health*

A vial of a precious, unstudied chemical sits on a shelf long, long forgotten, only to be almost tossed into the trash heap in the midst of a semiannual lab cleanup. What if it is thrown away? The world loses out on the first benzodiazepine, chlordiazepoxide, and possibly all of the modified forms to follow (including one of the most important medications of the twentieth century, Valium). Intertwined with chlordiazepoxide is the career of Leo H. Sternbach, the scientist who saved chlordiazepoxide from the trash heap. We will see how he fled a tumultuous situation in Europe for safety in the United States prior to ushering in a welcomed new era of treatment for those suffering from anxiety and created a slew of other popular medications along the way.

A Refugee Lands in New Jersey

Leo H. Sternbach was born in 1908 in Austria-Hungary (now Croatia), the son of a pharmacist. Chemistry and an inquisitive nature came naturally to him, and a lifelong love affair with the former stood on display as he disassembled artillery shells left over from World War I to make his own fireworks.[1] Dangerous work, but his nimble mind and hands prevented any major accidents.

Later in his childhood, Sternbach's father moved the family to Krakow, Poland, to open a pharmacy, with the younger Sternbach staying in the country to pursue graduate education. Leo Sternbach earned his PhD in organic chemistry in 1931 from the University of Krakow. His next steps had to be meticulously thought out, as Sternbach was of Jewish ancestry, increasing the need to carefully calculate his career steps to avoid the rise of the Nazi regime in the region at the time. In the years after graduation, Sternbach would work with future Nobel Prize winner Leopold Ruzicka at the Swiss Federal Institute in Zürich, Switzerland, just as the Nazi's ran roughshod over Poland. Many of the scientists Sternbach worked with at the University of Krakow would be sent to the Sachenshausen concentration camp.[2] Sternbach would quickly move on from the Swiss Federal Institute, as he felt his Jewish ancestry was preventing him from advancing, leading him to take a position at F. Hoffman-La Roche (now known colloquially as Hoffman-La Roche) in Basel, Switzerland, in 1940. His stay in Switzerland would not last much longer, as Hoffman-La Roche had the rare institutional foresight to move all of the Jewish individuals in the employ of the company and their families to their U.S. facilities.[3] As of June 22, 1941, Nutley, New Jersey—a far cry from Krakow and Basel—would be home to the ever-kind Leo Sternbach and his wife, Herta, with the couple leaning on each other throughout the difficult transition.[4]

Once settled in the United States in the early 1940s, Sternbach was given quite the difficult task by Hoffman-La Roche—find an industrial synthesis pathway for the natural product biotin. Biotin is better known as vitamin B7, with supplements containing the vitamin often touting its ability to bring out the beauty in hair, skin, and nails. A deficiency in biotin often leads to a thinning of the hair, scaly skin rashes, and increased breakage of nails.[5] Sternbach's work in creating an industrial-sized synthesis route for biotin is still hailed as one of his great achievements. This is due to the difficulty of the chemistry involved, much of it because of the number of chiral centers that must be protected en route to creating the final molecule.[6] After his work on biotin, Sternbach entered the world of drug design, aiding in the development of trimethaphan camsylate (brand name Arfonad), a ganglionic blocker that calmed the action of the heart during surgery.[7] After his work on Arfonad would come his career-defining task, but it all started with a much simpler problem, with Sternbach forgoing the simple and deliberately choosing a more difficult route. This route would bring the world a whole new class of medications, the anxiety-fighting benzodiazepines.

Hoffman-La Roche was not looking for anything groundbreaking when it tasked Sternbach with making a "me-too" version of the popular meprobamate. Meprobamate was a pharmaceutical wonder drug in the 1950s sold by Wallace Pharmaceuticals as the antianxiety drug Miltown, complete with Milton Berle as a spokesman (and patient). Berle gave himself the nickname "Uncle Miltown" on his variety show.[8] Hoffman-La Roche wanted Sternbach to do something simple—modify the structure of meprobamate just enough so that it was still active in fighting anxiety and, importantly, not infringe on Wallace Pharmaceuticals' rights.

Sternbach did not want to do the simple thing, however. It was not in his nature, as this was the same gentleman who successfully synthesized biotin just a few years earlier. Sternbach turned his mind to creating a whole new class of drugs with antianxiety benefits, recalling a series of chemical compounds he synthesized while researching in Poland during the 1930s. The compounds, benzheptoxdiazines, were intended to be synthetic dyes, but the chemical compounds were simply not good at being dyestuffs.[9] Still, Sternbach remembered the benzheptoxdiazines, thinking their structural features would lend themselves to an antianxiety application and provide many jumping-off points for synthesis. We will see Sternbach cling to what he was already familiar with throughout his career, with this familiarity paying immense dividends. Sternbach would synthesize a plethora of these benzheptoxdiazine synthetic dyes turned hopeful antianxiety drugs, including at least forty different versions stemming from hept-1,2,6-oxodiazine.[10] Time and time again on animal testing, however, the compounds proved fruitless. The journey to the first benzodiazepine was not without the involvement of luck, however.

A Bit of Luck Befalls a Two-Legged Rat

After two years without success in the form of creating a new antianxiety drug, Hoffman-La Roche reorganized and tasked Sternbach with developing new antibiotics. In 1957, a couple of years into the antibiotic project, Sternbach was conducting the ever-arduous yet necessary task of cleaning up his laboratory. While doing so, Sternbach came across one more derivative of hept-1,2,6-oxodiazine, coded Ro 5-0690, and sent it off for animal testing on a whim. Testing showed Ro 5-0690 to be on par with if not better than meprobamate. A structural analysis showed Ro 5-0690 to be a benzodiazepine and the first in a new class of drugs.[11] Hoffman-La Roche would lend Ro 5-0690 the name "chlordiazepoxide" prior to bringing the drug to market under its brand name.

One of the initial tests to determine a compound's effectiveness as an antianxiety medication involved sending dosed mice on a climb up a wire screen ramp. If the mice were successful in climbing the ramp, the compound was not as effective as already existing ones, but if the mice slid down the screen, the drug discovery team might have a winner on their hands. Not the most fun for the mice but also much more humane than many animal tests conducted through the years (or outlined in this book—I would much rather be drugged and slide down a ramp than have urine injected into my abdomen like the guinea pigs in John Cade's lithium salt trials). The traces of history show a single technician, Beryl Kappel, to have conducted many of these tests during a seven-year time period overlapping with the discovery of chlordiazepoxide at Hoffman-La Roche. Mrs. Kappel used meprobamate and chlorpromazine, two sedating antianxiety drugs that dominated the market at the time, as the standards by which she measured the compounds Hoffman-La Roche scientists brought to her.[12]

Sternbach hailed from a generation of chemists who considered themselves "two-legged rats," often trying their synthesized drugs on themselves. This is not completely unreasonable and without merit or justification, as one can tell the difference between a basic and an acidic compound by taste—bases typically taste bitter, while acids (think of the citric acid in a lemon) taste sour. However, Sternbach was not looking only to sample the flavor; he sought to determine whether the drugs he was synthesizing actually worked. The day he tried chlordiazepoxide, Sternbach recalled going home extremely tired, but that's better than what occurred in a separate instance with a still-unknown compound. Sternbach tried the compound, and in the hours that followed, his coworkers were forced to call his wife to take him home when his legs became unstable. Sternbach remained in bed for the following two days.[13]

There's a second narrative at play here that often circulates with the story of the "lucky" discovery of the first benzodiazepine, one where Sternbach begins work on antibiotics per instructions from Hoffman-La Roche but continues to seek out benzoheptoxdiazines that may serve as antianxiety drugs. In this version of the story, Sternbach intentionally develops and tests Ro 5-0690 but sits on the discovery for six months. He then uses the excuse of a periodic laboratory cleanup to explain away Ro 5-0690 as something they found during the cleanup and then sends it off for "official" testing.[14] The predominant view of the story remains the one in which Sternbach is looked on by the gods of luck during the mundane task of cleaning his laboratory, serving to instruct scientists to always be on the watch for overlooked opportunities wherein serendipity can rear its head.

The first clinical trials of chlordiazepoxide took place in 1958, centering on elderly individuals. Unfortunately, the dosage of chlordiazepoxide given to the patients is what we would now deem excessive, leaving the elderly trial participants heavily sedated and with slurred speech. Their anxiety was limited, yes, but heavy sedation and slurred speech are not side effects desired from a new drug you are trying to bring to market. Hoffman-La Roche convinced physicians in the Galveston, Texas, area to trial chlordiazepoxide but at considerably lower doses, and this second set of clinical trials was successful. By 1960, chlordiazepoxide was on the market—a blazingly fast movement from benchtop to clinical trials to patients, but this was sixty years of U.S. Food and Drug Administration red tape ago. En route to going to market, the awkward name chlordiazepoxide became transformed to Librium, a brand name given for the feeling of stability and equilibrium it provoked in patients.[15]

Chlordiazepoxide is what we now consider a long-acting benzodiazepine. The half-life of chlordiazepoxide is on the order of about a day, but its effects are felt much longer. This is because molecules of chlordiazepoxide are metabolized to the still actively sedating molecule desmethyldiazepam by the liver, with desmethyldiazepam having an effect for approximately 100 hours.[16] Chlordiazepoxide and other benzodiazepines act on the body by binding near the receptor sites of the neurotransmitter gamma-aminobutyric acid (or GABA in molecular shorthand) on neurons, and in doing so, they cause GABA molecules to become more likely to bind. When GABA binds at the receptor sites, it allows chloride ions to rush into the neuron and make it more negatively charged. This results in polarized neurons that have an uphill climb to flip from an overall negative charge to a positive one, making it harder to excite them. This effectively inhibits the response of the neuron, resulting in a depression of the central nervous system and a sedating effect.

Improving Chlordiazepoxide and the Creation of Valium

Seeking to ever challenge himself by improving his prior creations, Sternbach continued to tinker with the structure of chlordiazepoxide in hopes of making a more effective drug. He certainly succeeded. Within a matter of a few years, Sternbach created diazepam, better known by the brand name Valium (the name is taken from the Latin word *valere*, meaning "to be strong").[17] Valium replaced chlordiazepoxide almost immediately after coming on the market in 1963.

As diazepam (Valium) is a modification of chlordiazepoxide, you would probably guess that they share several structural similarities, and you would be right. Diazepam is actually a truncated form of chlordiazepoxide with minor atomic changes, with diazepam coming in under 285 daltons and chlordiazepoxide just shy of 300 daltons.

In the 1960s and 1970s, Valium sold more than any other prescription drug in the United States, with Valium hitting 2.8 billion tablets produced in 1987 alone.[18] In time, Valium too would be replaced in the same manner it replaced chlordiazepoxide, with fellow benzodiazepine Xanax (generic name alprazolam) taking the place reserved for it on many psychiatrists' and general practitioners' prescription pads. Sternbach played no role in the development of Xanax, as it was a product of the Upjohn Company.

Through the decades, the anxiety-fighting benefits of benzodiazepines have been overshadowed by withdrawal symptoms, dependency issues, and their abuse. Hints of benzodiazepine dependency began coming to the surface in the 1970s, but the class of drugs continued to be prescribed at an astronomical rate. These hints were solidified by studies of prisoners who were treated with high doses of chlordiazepoxide, but at the time, it was uncertain if the lower doses of chlordiazepoxide and its sister pharmaceutical diazepam used in everyday practice would elicit the same dependency.[19] Benzodiazepine dependency rears its ugly head as the body begins to rely on the small molecules to function, eliciting a rapid return of the initial anxieties that the benzodiazepines were used to allay when the patient ceases to take the drug. The onset of the returning symptoms is often worse than they were before treatment. Withdrawal due to abrupt discontinuation of taking benzodiazepines also leads to many sleepless nights thanks to the common onset of insomnia. These withdrawal symptoms can be lessened by slowly reducing the dose (i.e., tapering) of the benzodiazepine over the course of several weeks.[20]

Countering the dependency of those who need and rely on the benzodiazepines to conduct a life closer to normal is the abuse of benzodiazepines, the recreational use of the drugs without a prescription. Such use can also lead to withdrawal symptoms if the medications are taken for long enough. Particularly dangerous is the mixing of alcohol and benzodiazepines, as both depress the central nervous system and in doing so act as a force multiplier, with the possibility of respiratory depression, coma, and death constantly looming when the two are used simultaneously.[21]

Interestingly, benzodiazepines aid those suffering from alcohol withdrawal. Alcohol acts akin to benzodiazepines on GABA, but the sudden break from drinking alcohol can result in the rapid overactivity of neurons. Benzodiazepines are used to slow this overactivity, with the benzodiazepines slowly tapered over time. Both diazepam and chlordiazepoxide are used to treat alcohol withdrawal, as they are longer-acting benzodiazepines and can calm the neuronal storm caused by alcohol cessation thanks to their stability.[22]

Sternbach's Enduring Legacy

Despite the negatives associated with benzodiazepines, Sternbach's contributions of Valium and chlordiazepoxide alone would easily make him one of the most successful drug discoverers of the twentieth century. However, Sternbach did not stop with the pair. In addition to Valium and chlordiazepoxide, Sternbach also played a role in the development of the antianxiety drug Klonopin, the insomnia drugs Mogadon and Dalmane, the peptic ulcer drug Quarzan, and the unfortunately often-abused Rohypnol. Many of these are within the benzodiazepine class, showing Sternbach's propensity to continue to tinker with earlier successes in hopes of finding something new.[23] In all, Sternbach was named as a creator in 241 patents by the end of his career.[24]

Sternbach's work transformed Hoffman-La Roche into the pharmaceutical giant it is today. At one point in 1994, work emanating from Sternbach's patents accounted for 28 percent of Hoffman-La Roche's worldwide pharmaceutical sales.[25] While more than a quarter of the financial bottom line of the company stemmed from the brain-and-chemistry know-how of Leo Sternbach, in keeping with many of the tales of discovery throughout the history of pharmaceuticals, Sternbach profited a mere $1 per patent—meaning a single dollar each for the discovery of chlordiazepoxide and diazepam—as he received a salary through Hoffman-La Roche, and, per his contract, the company owned all his work-related discoveries.[26]

Hoffman-La Roche did single out Sternbach several times for individual awards and prizes, including monetary ones, and Sternbach did ascend through the company's hierarchy on the back of his work, retiring in 1973 as director of medicinal chemistry. The kindhearted Sternbach continued a relationship with Hoffman-La Roche as a consultant, with the company no doubt getting more than its fair share of the bargain as Sternbach continued to come to the office until 2003 at the age of ninety-five.[27] The reason he stopped? A move with his wife, Herta, from New Jersey to North Carolina to be closer to Daniel, one of his two sons, who followed in his father's footsteps albeit it at rival pharmaceutical company GlaxoSmithKline.[28] One cannot fault Daniel, however, as it would no doubt be an insurmountable task to escape from his father's long-reaching shadow, even if Leo Sternbach saw himself as a two-legged rat.

CHAPTER 8

Nitrous Oxide

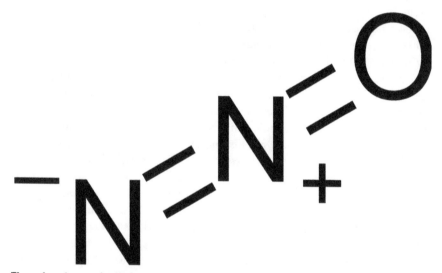

The structure of nitrous oxide, better known as laughing gas. *Structure generated by author in ChemDoodle v 11.5.0, using information supplied by the National Institutes of Health*

First synthesized in 1772, it would be more than 100 years until dentists and physicians used nitrous oxide for the sedating and pain-relieving benefits for which we are now accustomed to its being utilized. En route, the gas passed through the hands of two of the most accomplished scientists of the era, Joseph Priestly and Humphry Davy. Once out of their hands, nitrous oxide became a bit of a parlor trick. Nitrous oxide was used to draw large crowds at traveling exhibitions, with the audience welcome to give the gas a try as long as they were willing to pay a fee. However, one fateful night at an exhibition laid the groundwork for its future use in medicine and the destruction of one man's life.

Joseph Priestly: Polymath, Rabble-Rouser, and Discoverer of Nitrous Oxide

The son of a cloth finisher—an arcane title for someone who used chemical processes in the textile industry—Joseph Priestly was born in 1733. It appears the textile business held too much of a concern in the heart of his father, as he was not the sort to raise a family once becoming a widower. Shortly after Priestly's birth mother died, the young Joseph was sent to live with his aunt at the age of six, Mrs. Keighley, who would be every bit the mother in Joseph's life despite losing her husband shortly after Joseph came to live with them. She made the young boy her "project" and molded and shaped the young man by imparting her own nonconformist political and religious views. In stark contrast with his father, the recently widowed Keighley gave Joseph a permanent home by adopting him in 1742.[1] Priestly bore a fitting name, for he shined brightly in this period before the schism between religion and science. Keighley also oversaw Priestly's education, spurring on a love of learning that would lead him down the path to become one of the eighteenth century's great polymaths, excelling in both religious and scientific studies while learning nine languages along the way.

This is made all the more astonishing by the fact that access to the finest schools did not come as easily to Priestly (if at all). Priestly's doting aunt raised him with her religious beliefs, leading Priestly to be a Dissenter, someone who did not side with the beliefs of the Church of England. As a Dissenter, this limited his educational opportunities, as the doors of Oxford and Cambridge would be closed to him. As a teenager, Joseph attended the Dissenting Academy at Daventry, and afterward, Priestly would take a ministerial position overseeing a small church in Suffolk, England, at the age of twenty-two. Despite his intellectual prowess, money was scarce, as the church paid him the sum of £30 for an entire year's labor, leading Priestly to operate a small school from the church to supplement his income. On top of this, Priestly spoke with a natural stammer, forcing him to develop a deliberate yet relaxed style for his sermons and lectures. While teaching, he began writing his first book, a grand demonstration of the breadth and depth of his knowledge, a grammar text titled *The Rudiments of English Grammar*, in 1761. The book was quite popular, seeing use well into the next century. Priestly followed this tome with one completely different after befriending Benjamin Franklin while in London and becoming fascinated by Franklin's work on the nature of electricity. Building on Franklin's seminal studies, Priestly carried out a number of his own studies and experiments, publishing them as *The History and Present State of Electricity* in 1767. While creating drawings for a second edition of this work, Priestly discovered that caoutchouc—natural rubber, a new introduction to England in the eighteenth century—

would erase pencil lines, replacing the customary use of bread crumbs to perform the same task. Around this time Priestly also created the precursor of modern soft drinks, Pyrmont Water, by bubbling carbon dioxide (ever in supply thanks to the brewery next door to his house) into water from a local mineral springs to create carbonated water.[2] Spurred on by the interesting effects of carbon dioxide on water, Priestly would move on to a study of various gases (albeit forever calling them "airs" instead of our modern parlance of gases) in what would prove to be the most fruitful scientific work of his career. To record these observations, Priestly published a number of volumes of *Experiments and Observations on Different Kinds of Air*, with the first coming in 1775.[3]

The discovery and isolation of oxygen would be Joseph Priestly's most important achievement and the core of his studies of various gases, or "airs," as he deemed them. To study gases, Priestly used a twelve-inch piece of glass he called a burning lens, so called because it allowed him to concentrate the light of the sun to heat an object within its sight. One day in 1774, Priestly placed a gorgeous red-orange sample of what we now know to be mercuric oxide in the path of his burning lens and collected the gas that arose as the mercuric oxide decomposed. Priestly became fascinated with this gas, which he called "dephlogisticated air," and marveled in its ability to keep a flame burning strongly for an extended period of time.[4] The gas Priestly called dephlogisticated air? Oxygen.

Priestly conducted animal studies with what he called dephlogisticated air and what we now know as oxygen using a cleverly devised system to keep any other airs from contaminating his experiments. To maintain an uncanny degree of experimental fidelity, Priestly inverted a glass experimental chamber and filled the surrounding outer container with mercury, ensuring that the air inside the inverted container would neither be added to nor removed thanks to the density of mercury. Using mercury in such a manner is frowned on today because we now know mercury exposure to be highly toxic, as it inhibits a number of selenium-dependent proteins in the body that allow the cycling of vitamins C and E and a number of antioxidant protective features. Despite consistent exposure to mercury during the course of his gas experiments wherein he would discover not only oxygen and nitrous oxide but also ammonia, sulfur dioxide, and silicon tetrafluoride, we do not have any record of sickness on the part of Priestly, who would live to the reasonably old age of seventy.

In his experiments, Priestly collected data on three interests: how long and how well a flame would burn in the gas he had created as well as the length of time an animal—most often a mouse—would survive in the gas. With oxygen showing a propensity for maintaining a vibrant flame, Priestly turned his eye to the latter. In his animal studies, Priestly placed a mouse in the inverted glass container containing his dephlogisticated air and observed that the mouse lived for an hour. This was four to five times longer than a mouse would live in control experiments, wherein the glass container contains only ordinary room air.[5]

Priestly created nitrous oxide, which he called "dephlogisticated nitrous air," through a slow, arduous process in which he placed strips of zinc in diluted nitric acid and collected the resulting gas (a gas often laden with impurities).[6] This method would soon be usurped for a much more efficient reaction in which ammonium nitrate is heated and a pure gas collected.[7] When it came time to test nitrous oxide after he discovered it two years earlier in 1772, Priestly noticed that a burning flame was greatly increased in the presence of the gas—much like with oxygen—but that the mice he subjected to the gas quickly died (or so he thought, as we will see later). The discovery of nitrous oxide and his limited animal experiments with the gas would be the extent of Priestly's studies on nitrous oxide.

The latter portion of Priestly's life would take on a distinct political nature, although he himself never held public office. Priestly supported both the American and French revolutions—not the most earth-shattering opinions today, but for a public figure entrenched in the social stratosphere of Birmingham, England, this was not the most popular stand to take. Priestly railed against the establishment of his country and its predominant religious views and stood as an ardent supporter of the uprisings. Priestly called his words "grains of gunpowder which would one day explode under the Anglican Church."[8] His support of the French Revolution—a violent overturning of the social strata that many in power in England feared would spread to its shores—led to the Priestly Riots, wherein an angry mob took to the streets on July 14, 1791. This was the second anniversary of Bastille Day, with rioters destroying Priestly's laboratory, his home, and the homes of other prominent local Dissenters along with their house of worship.[9] After the Priestly Riots, Joseph Priestly and his family left Birmingham, as would be natural for anyone living through a riot named after them. The family spent a few uneasy years in North East London before leaving England altogether for the shores of the nascent United States of America in 1794, settling in Northumberland, Pennsylvania.[10] His first aim was to create a religious community for like-minded individuals in North America, but this endeavor failed. While in Pennsylvania, Priestly turned down a position as the chair of chemistry at the University of Pennsylvania, a school founded by Benjamin Franklin. Instead, Priestly founded the first Unitarian Church in the United States.

Throughout his life, Priestly collected several esteemed acquaintances, calling Erasmus Darwin (grandfather of Charles Darwin and mentor of the formerly mentioned William Withering), Benjamin Franklin, and Thomas Jefferson friends at one point or another. Erasmus Darwin was a fellow member of the Lunar Society, a Birmingham meeting of intellectuals that took place on Monday nights near a full moon, allowing the members to talk into the early morning and still walk home safely accompanied by lunar light.[11]

Despite his successes and progressive political views regarding the American and French revolutions, Priestly held fast to some quirky and antiquated scientific beliefs. He clung to the phlogiston theory—a thought process explaining what we now know as chemical reactions as events in which transfers of a fire-like element to and from objects occur. Phlogiston theory is a little hard to wrap our minds around due to our modern understanding of chemical reactions, but in this theory, an object that burned readily was said to be rich in phlogiston, with the phlogiston released as the object burned. A portion of Priestly's belief in phlogiston theory possibly stems from a lack of training in mathematics and physics. Also, Priestly never considers concepts like the law of conservation of mass, a concept accepted by most of the scientific world at this point in history. Joseph Priestly never departed from the theory of phlogiston, holding firm to the belief until the day he died. In addition to phlogiston theory, Priestly, as noted a couple of times so far, never referred to a gas as a gas per se, instead choosing to refer to nitrous oxide, oxygen, and other gases simply as "airs."

While Priestly held some unusual beliefs, he was no doubt a brilliant mind and stood firm for what he believed in. Although he never applied nitrous oxide to the human body in hopes of discovering a medical use, our next key role player, a young Humphry Davy, seized the opportunity and discovered a realm of possibilities for the gas, all before the age of twenty-three.

Humphry Davy's Rapid Ascent and Self-Experiments with Nitrous Oxide

Born to a seldom-employed woodcarver and his wife in the coastal town of Penzance in 1778, Humphry Davy would be but a toddler as Joseph Priestly was fighting for his life amidst riotous mobs in the city of Birmingham 300 kilometers north.[12] Like Priestly, Davy suffered the death of a parent at a young age, calling an end to his costly formal education and nudging him into an apprenticeship with a prominent surgeon in Penzance, John Bingham Borlase, at the age of fifteen. A career as a physician appeared to lie ahead of him; however, Davy lacked the drive one would typically associate with the profession. He is described as idle and inattentive during this period, spending most of his time playing pool or wandering by the coast. At eighteen, he decided this life would no longer be fitting for him. Davy abandoned a path to become a physician and instead formed his own curriculum of self-instruction, a curriculum steeped in chemistry and experimentation. Davy was ever resourceful once he found his passion, and this is demonstrated by his prize possession at this point in his life—a glass enema syringe he received from a shipwrecked French surgeon. Why such value on a syringe? The young Davy was able to use it to create an air pump and further his chemistry experiments. Mere months into his study, Davy devised that Antoine Lavoisier, the foremost mind in the field of chemistry during the eighteenth century, failed to account for the action of light in producing heat and its presence as a tangible part of chemical reactions.[13]

Davy became enamored with nitrous oxide due to controversial claims being made about the gas by New York–based professor Samuel Latham Mitchill. Mitchill convinced himself and others that Priestly's dephlogisticated nitrous air—thankfully renamed in short time by Davy as nitrous oxide—was the source of a miasma, a "bad" air behind such diseases as the plague. Miasma theory was not uncommon in this period, but Davy devised a set of experiments to disprove Mitchill's assumptions. Davy placed mice in containers filled with dephlogisticated nitrous air and breathed the gas in. He also exposed open wounds to the gas to see if they would become infected, going so far as to place strips of meat into a container of the gas to observe whether they putrefied faster in the presence of dephlogisticated nitrous air. Davy soundly defeated Mitchill's theories and in doing so trod on a new path of study that would define his early career, the examination of gases, chiefly nitrous oxide. To gain further understanding of nitrous oxide, Davy simply repeated many of Priestly's experiments with the gas, yet in doing so, he found slightly more nuanced results. Instead of determining that small animals die in its presence, Davy noticed the animals were merely sedated if exposed to the gas for a few minutes at a time, and once the animals in question (most often mice) were removed from the experimental device, they regained consciousness.[14] All of these experiments were conducted at a very young age. One could say Davy aimed to make up for lost time, for by the age of nineteen, he stepped into the role of superintendent of the Bristol Pneumatic Institute, a hospital established to use various gases to treat perplexing diseases of the day, such as dropsy, asthma, consumption, "obstinate venereal complaints," and scrofula—a series of external tuberculosis complications and lesions also known as the King's Evil.[15] Why the King's Evil? This more ominous name stemmed from a belief that merely the glancing touch of your regional monarch could cure the disease.[16] Whether the Bristol Pneumatic Institute was successful in curing any of these diseases is certainly questionable, but the position gave Davy time to carry out more experiments on nitrous oxide and write. Two years later, he would compose the foremost treatise of the era on nitrous oxide, *Researches Chemical and Philosophical, Chiefly Concerning Nitrous Oxide and Its Respiration*, a nearly 600-page book detailing the preparation of nitrous oxide and similar gases as well as the effect of the nitrous oxide on animals.[17]

Unlike Priestly, Davy focused his experimentation on nitrous oxide instead of jumping from "air" to "air" as Priestly did. Davy was so fond of the gas that he even used it recreationally for its mood-lifting effects, declaring it left him with a sense of "pleasurable delirium."[18] He also self-experimented, as was typical in the time of these early chemonauts, but Davy's personal experiments were often a little more fun than most. The day before Christmas Eve in 1799, Davy deemed it necessary to test whether nitrous oxide calmed the effects of alcohol. To do so, Davy drank an entire bottle of wine in eight minutes before inhaling a copious amount of nitrous oxide, with his notes suggesting the gas served to ameliorate the effects of his hangover. Davy found that headaches often ceased after imbibing the gas, leading him to give nitrous oxide a try when opportunity arose in the form of an infected wisdom tooth. The relief from pain gained through the use of nitrous oxide led him to suggest in a notation made in 1800 that the gas could be of use in surgery.[19] The notes were not circulated well and as such led to a half-century wait for physicians aiming to relieve the anguish of patients going under the knife.

Davy's Role in Discovering Eight Elements and His Greatest Discovery, Michael Faraday

Davy would cease experimenting with nitrous oxide altogether in 1801, turning his eye to what is now known as the field of electrochemistry and in doing so becoming a pioneer in elemental discovery. Davy initiated the second half of his career at the age of only twenty-three, a period when most scientifically minded adults today are just finishing college. At this time, Davy began conducting the electrolytic experiments that would lead him to discover no fewer than eight of the elements on the periodic table—sodium, potassium, barium, boron, calcium, chlorine, magnesium, and strontium. Despite these achievements, Davy would often remark that his young assistant, Michael Faraday, was his greatest discovery. Faraday would go on to far surpass the work of Humphry Davy thanks to his advances in electromagnetism. The story of Faraday and Davy's initial meeting comes in many different forms, but what cannot be disputed is Davy taking on Faraday as an assistant shortly after a lab accident wherein the detonation of nitrogen trichloride significantly damaged Davy's eyes.[20] Davy needed an assistant, as his own fame granted the elemental master the opportunity to leave England and travel across Europe, including France. This was quite the unusual occurrence at the time, as England and France were deeply entrenched in the Napoleonic Wars. Napoleon established a blockade of the English Channel, with Davy and Faraday crossing the blockade on a prisoner exchange ship to make their way to Europe in 1813.[21] The duo would spend two years studying and interacting with landlocked scientists in the region, aiding when they could, aid that included helping identify the element iodine.

On returning from Europe, Davy turned his eye to altruistic endeavors, including solving a problem threatening the life of every miner in England at the time. The problem? The use of open flames to light the way as miners trudged through cavern after cavern removing coal and other necessities for the burgeoning industrial world. These open flames often came in contact with pockets of methane gas and led to deadly accidents. One such incident particularly struck Davy, the 1812 Felling Colliery disaster, which claimed the lives of ninety-two men and young boys. To eliminate the use of open flames, Davy created an oil-based lamp enclosed in a wire mesh, a lamp that could light the way while separating the heat source from pockets of flammable gas. This invention worked as the mesh absorbed the majority of the latent heat from the flame, while the tiny openings in the mesh still allowed a substantial amount of light to shine ahead in the mines. The creation quickly became known as the Davy Lamp. Davy, however, refused to take out a patent on the device due to its lifesaving application. This would be the last great achievement of Davy's career, with the scholar dying from complications following a stroke—a stroke likely due in some part to chemical exposure throughout the course of his career—at the age of fifty.[22]

Laughing Gas Exhibitions and the Tragic Story of Horace Wells

Although Davy succeeded in popularizing nitrous oxide, he did not usher it into the mainstream medical use we see today, as his notes suggesting a surgical application failed to circulate. The necessary connection would come through an unusual source—traveling "laughing gas" shows, events extolling the odd behavior that came about when unwitting participants inhaled nitrous oxide for the first time. These shows were quite popular, with Samuel Colt touring through the United States and Canada giving laughing gas exhibitions. Colt eventually used the money earned to fund the creation of handgun prototypes with a rotating cylinder, the foundation of his armament empire.[23] The showman who played a role in getting nitrous oxide to the medical community would not be Samuel Colt, however, but one of his contemporaries in the following decade, a Mr. Gardner Quincy Colton.

The broadside for the December 10, 1844, laughing gas exhibition in Hartford, Connecticut, is a wonderful piece of nineteenth-century marketing and showmanship. In it, Gardner Quincy Colton speaks of the forty gallons of nitrous oxide to be distributed at the show, with anyone paying the twenty-five-cent entry fee (roughly $10 today) welcome to try the gas. Colton proclaims that eight strong men will be on guard duty during the show to protect those imbibing the gas from injuring themselves and that "probably no one will attempt to fight." Probably. Colton notes that 4,000 people paid to attend a show of his in New York earlier in the year, adding that those who inhale the gas once are always "anxious to inhale it a second time." At these exhibitions, Colton would give a short lecture to introduce the gas and give an overview of its scientific properties before opening up the forty-gallon supply to the audience. Colton— ever the showman—was even so kind as to offer a "ladies only" hour from noon to 1 p.m. on the eve of the exhibition so that women could try it for free in the absence of men.[24]

So, apart from posterity and comedic context, why is this broadside from the December 10, 1844, show so important? In attendance at this exhibition was Dr. Horace Wells, a local physician who took much joy and interest in watching those taking samples of the gas. We do not know if Wells imbibed nitrous oxide on that evening, although I would argue that it is unlikely. Wells paid close attention to one individual who did imbibe, a Mr. Sam Cooley. Cooley, after taking his drag from the rubber bags in which Colton kept the nitrous oxide contained, proceeded to run from the stage and pursue a blameless attendee in the crowd with the intent of pummeling him. The attendee, who did not know Cooley, fled. Cooley quickly followed, jumping over seats and a settee before running down the aisles of Hartford's Union Hall. Then, almost as if leaving a trance, Sam Cooley stopped and sat down in a seat near Dr. Horace Wells. Dr. Wells noticed Cooley was bleeding from the leg, an injury sustained while jumping over the settee in the hall. Cooley had no recollection of the wound, nor did he feel any pain until minutes later. Wells, nimble of mind, put together a connection between Colton's laughing gas and the absence of pain and planned to test his hypothesis the next day.[25]

How would Dr. Wells test whether nitrous oxide truly alleviated pain? By following the era's modus operandi of self-experimentation and by carrying out a painful medical procedure on himself, of course. Wells contacted the previous night's laughing gas exhibitor, Gardner Quincy Colton, and asked Colton to supply him with nitrous oxide while a friend, Dr. Riggs, extracted a molar from Dr. Wells's jaw. The test was a success, as no pain was felt, and Riggs and Wells went on to successfully repeat the procedure in several more tooth extractions.

With nitrous oxide demonstrated to be an analgesic in minor surgery, Dr. Horace Wells believed he had a phenomenal achievement and had to communicate his findings to the medical community. With a number of repeated procedures under his belt, Wells contacted Boston's Massachusetts General Hospital in 1845 and offered to give a demonstration of nitrous oxide's powers. Massachusetts General Hospital is a teaching hospital long affiliated with Harvard Medical School. This was the chance of a lifetime—center stage at one of the most prestigious hospitals on North American shores—and the operating gallery was packed with attendees. As the tooth extraction was performed in the presence of nitrous oxide, the patient unfortunately cried out in pain, leading the audience of medical students to boo and then begin chanting "humbug" and laughing.[26] Wells, now distraught, claimed that the nitrous oxide was discontinued too early and that this led to the patient's suffering. Wells received no mercy from the crowd or the medical establishment of Massachusetts General Hospital.

Sadly, this singular event sent Horace Wells's life spiraling out of control. Wells quit dentistry, instead becoming a traveling salesman, selling everything from canaries to bath items. Shortly thereafter, he became addicted to chloroform, leaving his family behind in Hartford and moving by himself to New York City. On an evening in January 1848, Wells exited his abode and threw sulfuric acid on two prostitutes, splattering one in the face and neck with the acid and sending her to the hospital. For this dreadful and unprovoked attack, one likely brought on through his use of chloroform, Wells was sent to Manhattan's Tombs prison, but due to his mental state, his stay was not long. Early in the morning of January 24, 1848, an ashamed and spiraling Wells found a shaving razor and committed suicide by slashing his femoral artery.[27]

In the years that followed, William Morton, an associate of Dr. Wells with firsthand knowledge of his work with nitrous oxide, partnered with a Boston-area physician Charles Jackson to successfully combine nitrous oxide with perfumed ether to yield an anesthetic effect. The duo was triumphant where Wells was not, convincing the faculty of the Massachusetts General Hospital to make use of their mix of gases in surgery. It is often quoted that Massachusetts General Hospital physician John Collins Warren proclaimed, "Gentlemen, this is no humbug!" at the successful demonstration by Morton and Jackson, but this parallel use of "humbug" appears to be a fanciful construction of history.[28] Morton aimed to gain full control of the use of these gases in medical applications, patenting the mixture and even going so far as to seeking out the heirs of Horace Wells to purchase the rights to any patents they might hold.[29] Morton's administration of ether in the mixture benefited him through one of its basic chemical characteristics—ether is a liquid at room temperature and thus could be transported easily. Nitrous oxide, on the other hand, had to be synthesized close to the time of use and carried in often-leaky "rubber bags." Ether was also readily available through industrial processes, making its use all the more accessible.[30]

But use of nitrous oxide alone would not be dead. Along with Morton, Gardner Quincy Colton, the carnival barker we encountered earlier, would reap the early benefits of the use of nitrous oxide in dental anesthesia. In 1863, he established the Colton Dental Association, profiting immensely from the application of nitrous oxide to tens of thousands of patients over the course of the next several years.[31] Today, nary a pediatric appointment or extended adult dental procedure occurs without the use of nitrous oxide, with the gas mixed with oxygen to prevent complications. Despite the natural draw of a pain-free tooth extraction or surgery, both surgeons and patients were often reticent. An 1846 report of a patient undergoing a tooth extraction shows the patient refusing anesthesia for fear that he would wake up and learn the dentist had removed all of his teeth. The Russian surgeon Nikolai Pirogoff found it "repugnant" to operate on patients subjected to anesthesia in the nineteenth century, growing accustomed to the cries for help and screams and viewing them as useful in his surgical investigations. Pirogoff later changed his opinion, thankfully, but only after growing to view the use of anesthesia as a benefit to the surgeon, increasing the types of operations they could comfortably perform.[32]

Use (and Abuse) of Nitrous Oxide Today

In addition to dental uses, nitrous oxide grew to see use in more complicated medical procedures. By the 1930s, nitrous oxide was the primary means to alleviate the pains of childbirth. Today, nitrous oxide sees use in sedating pediatric patients for minor surgical procedures, as it removes the necessity for intravenous anesthesia administration or allows for easier application of an IV once the child is sedated. A pleasant smell like bubble gum or grapes can be added to the nitrous oxide to make a child more receptive to putting on a mask and breathing deeply. Nitrous oxide may have an expanded use in general surgery in the future as well. A recent study shows that giving a patient nitrous oxide in the middle of a surgery decreases the experience of chronic postsurgical pain.[33] A more curious use of nitrous oxide comes in its use in treatment-resistant depression, where small studies have shown nitrous oxide to yield immediate as well as multiday antidepressant effects in patients who have failed to find relief with two or more previous forms of pharmacological treatment.[34]

Nitrous oxide is a tiny molecule, roughly 44 daltons, and the smallest thus far of any we've encountered in this book save for the lithium ion. The gas is colorless and has a faint and difficult-to-describe but not noxious smell. We do not know the exact biochemical mechanism by which nitrous oxide provides its analgesic and anxiolytic effects, and it may very well be that multiple mechanisms are at play. The pain relief observed with administration of nitrous oxide stems from a release of opioid peptides held in the neurons, but the exact manner explaining how the binding of a nitrous oxide molecule produces this effect is unknown. The anxiolytic and mild sedation felt under the influence of nitrous oxide likely comes in a similar manner by which benzodiazepines exhibit their antianxiety action: activating the GABA receptor, which results in negatively charged chlorine ions entering the neurons and altering their state and decreasing their excitability. Finally, nitrous oxide's anesthetic properties (likely its most important) appear to be governed by the inhibition of N-methyl-D-aspartate (NMDA)–type glutamate receptors, thus taking their excitatory function out of the equation for the time during which nitrous oxide is applied to a patient.[35]

Despite the positives we have outlined, there are some negatives concurrent with the continued use and availability of nitrous oxide. Nitrous oxide is a greenhouse gas, with the gas accounting for 6 percent of all man-made greenhouse gas emissions in the United States. Although the percentage may be small, nitrous oxide is very efficient at trapping heat—nearly 300 times better than carbon dioxide. This magnifies its detrimental effects, especially considering that the gas was essentially nonexistent in the atmosphere until a little more than 200 years ago. Additionally, nitrous oxide undergoes a series of reactions in the presence of light and oxygen when the gas is in the middle stratosphere, resulting in a molecule of nitrous oxide being broken down into two molecules of nitric oxide. These two molecules of nitrous oxide are then able to react with individual molecules of ozone to produce nitrous oxide and oxygen again, leading to the piecewise deterioration of the ozone layer.[36]

The access to nitrous oxide and the "pleasurable delirium" it brings per Humphry Davy's observation also opens the chemical up to being abused as a recreational drug, particularly in the United States and the United Kingdom. The access, overall legality, low cost, and absence of nitrous oxide from most routine toxicology screens come together to bolster its popularity. The effects of nitrous oxide also dissipate rather quickly. Nitrous oxide inhalation is not without its detrimental effects, however, many of which are quite dangerous and alarming. Inhaling directly from a pierced metal canister of nitrous oxide, called a whippet, is particularly dangerous, as the rapid increase in volume that occurs as pressurized nitrous oxide exits its former home in the steel canister causes the gas to spew forth at an extremely low temperature. This quick rush of gas often leaves the user with severe frostbite on the face and mouth. Inhaling from a concentrated supply also opens up the user to momentary hypoxia as the nitrous oxide rushing into the nose and mouth displaces oxygen and could lead to a seizure or, in dire cases, arrhythmia, leading to a heart attack. Even so, not all of nitrous oxide's detrimental impacts happen at the point of use. Continued abuse of nitrous oxide over a long period of time leads to the inhibition of the enzyme methionine synthetase and the inactivation of vitamin B12 in the body, the latter leading to numbness and a decrease in positional awareness of the hands and feet as well as a generalized degeneration of the nervous system.[37]

The story of nitrous oxide is a tale of many hands. What started as an abandoned discovery by the ever-busy Joseph Priestly gained prominence in the hands of Humphry Davy before finding an application in the arms of Wells (albeit with a tragic end for the good doctor). Of the trio, likely only Davy could have foretold its possible abuse. Despite the avenues of abuse, the dental and medical applications are substantial, and without them, many of our routine surgeries and dentist visits would be far more painful and anxiety producing.

CHAPTER 9

Nitrogen Mustards

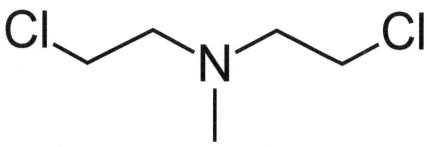

Chlormethine, one of the nitrogen mustards. *Structure generated by author in ChemDoodle v 11.5.0, using information supplied by the Royal Society of Chemistry*

The term "nitrogen mustard" likely brings one thing to your mind—the horrors of mustard gas. This terror of World War I led to untold physical and mental trauma as soldiers on both sides of the battle stood endlessly at heightened states of readiness should shells containing one of the first true weapons of mass destruction fly in their general direction. While the cancer-fighting nitrogen mustard chlormethine never saw use as an agent of war, it is forever connected to mustard gas. Just how did a weapon of mass destruction from the early twentieth century pave the way for chlormethine and the first form of chemotherapy?

Chemical Weapons in World War I

Frederick Guthrie of the London Institute of Physics is believed to be the first to synthesize sulfur mustard, doing so in 1860. Per the standard of the time and like Humphry Davy and others before him, he took to testing it on himself, but instead of enjoying the euphoric effects of nitrous oxide like Davy, Guthrie suffered the blistering that would come to be associated with mustard gas exposure.[1]

The effects Guthrie suffered rolled around in the back of the minds of devious chemists for decades, however, before it would see its unfortunate military use.

While mustard gas is synonymous with chemical warfare, mustard gas was not the first gas used in modern warfare, as there are at least two that preceded it. The first attempt at weaponizing a gas is often looked back on as a failure, in part due to the freezing temperatures in which it was dispersed. At the Battle of Bolimov in January 1915, German troops fired shells loaded not with explosive munitions but instead with the small molecule xylyl bromide, known as "Substance T," to those on one side of the battlefield that day. The shells penetrated Russian lines and released the noxious gas. Due to the cold weather that day, the xylyl bromide gas froze in midair, leading to the attack being seen as a failure despite the nearly 1,000 Russian deaths associated with this first use.[2]

Although the use of xylyl bromide was a surprise to the Russians, the potential for chemicals to be used in warfare was not. The Hague Declaration of 1899 and the Hague Convention of 1907 took steps to forbid the use of "poisonous weapons" in warfare, but, as we will see, this rebuke did not stand up.[3] It is suspected that the German army attempted to use chemical weapons prior to the release of xylyl bromide but with no success. The idea of using chemical weapons existed as far back as the fifteenth century, when inventor and painter Leonardo da Vinci proposed loading empty ballistic shells with chemicals to fire at attacking ships.[4]

The second use of a gas in warfare would succeed where the first did not and solidify chemical agents as weapons of mass destruction. This step into a new world of warfare took place on April 22, 1915, at the Second Battle of Ypres. Ypres is an ancient town in Belgium mentioned in Chaucer's *The Canterbury Tales*, a town that would unfortunately become better known as the site of many a conflict between Allied and German forces.[5] Within ten minutes, 160 tons of chlorine gas flowed from 6,000 steel cylinders installed by the Germans three weeks earlier. Why weeks earlier? The Germans waited for the optimal moment to maximize the devastation in their hopes of bringing an end to an enduring struggle in the trenches at Ypres. According to plan, the gas flowed directly toward the Allied line on wind currents carefully studied by German meteorologists. History leaves us with the perspective of a British soldier stationed at the rear of the line, who saw "greenish-gray" clouds float toward the front lines before the gas turned yellow and shriveled the plant life surrounding his fellow soldiers. The worst was yet to come, as the soldier added, "Then there staggered into our midst French soldiers, blinded, coughing, chests heaving, faces an ugly purple color." The French and Algerian forces on the front lines suffered the brunt of this new form of attack, killing 1,000 soldiers within minutes and wounding an additional 4,000.[6] The latter would be the true power of these gaseous chemical weapons—not the ability to kill but the ability to remove large swaths of the opposing forces from the battlefield in a single attack. On average, a soldier exposed to chlorine gas would spend sixty days in recovery (if surviving at all), placing an immense toll on an already taxed military infrastructure.[7]

The mind behind the chlorine gas plot was future Nobel Prize winner Fritz Haber, although he would win the 1918 award for much more altruistic reasons—the development of the Haber-Bosch process, a method of producing ammonia for use in fertilizer. Despite the success of the chlorine gas attacks, Haber did suffer directly for his role. On the eve of the deployment of the gas, his wife, Clara Immerwahr, a fellow PhD in chemistry, was overcome by guilt due to her knowledge of the impending slaughter. This guilt, combined with pressures in their marriage at the time, led Dr. Immerwahr to take her own life.[8]

The idea of a future Nobel Prize winner like Fritz Haber being pulled from the ivory tower to work on such a ghastly scientific endeavor shows the intermingling of academia and the chemical industry with the military in the total war approach taken during World War I. Fritz Haber was far from the only major scientist charmed by the war effort, with another future Nobel Prize winner in chemistry, Walter Nernst, pushing for the use of chemicals as weapons of war as early as 1914.[9] Regardless of his previous research affiliations, Fritz Haber certainly took to his military role, looking almost like a super-villain in his uniform, with his visage no doubt playing a role in the design of Baron Wolfgang von Strucker by Stan Lee and Jack Kirby in the early 1960s heyday of Marvel Comics.

In time, chlorine gas attacks would become common, with soldiers issued protective gear and drilled repeatedly so as to practice getting on their lifesaving protective attire at a moment's notice. Why? It was not uncommon for the sirens announcing a chemical attack to come at any hour of the day or night. As the protective gear became better and the troops more acclimated to putting it on, early chemical weapons like chlorine gas became less effective, necessitating the development of new ones. Here the story of chlormethine and nitrogen mustards begins.

It would take two years for mustard gas to see its first use in World War I, when, with the impending reality settling in that the war may be hopelessly lost, the Germans unleashed the "king of battle gases" in July 1917.[10] By this time, the United States counted itself as an active member of the Allied forces, standing firmly beside France, Great Britain, and a legion of others to combat the expansion of the Central Powers. Mustard gas was distinctly different than the chlorine gas used before—instead of targeting the lungs, mustard gas acted as a vesicant. What is a vesicant? A chemical causing blisters on exposed areas as well as condemning those exposed for long time periods with permanent ocular damage.

The Germans called mustard gas "yellow cross" because of the color and shape used to designate shells that held the irritating oil. The British and Allied forces often called it "HS" or "Hun stuff," a slang name taken from a pejorative for the German army.[11] Mustard gas, however, was developed under the code name LOST, an acronym for the scientists who created it—Wilhelm Lommel and Wilhelm Steinkopf (LOmmel and STeinkopf).[12] Their creation was sulfur mustard—which stands as a more refined term for mustard gas—a molecule that differs from the eventual chemotherapy drug but provided the structural basis for its origin. I say refined here, as mustard gas is not a gas at all but an oil that smells of mustard seeds. The oil is vaporized by explosive shells in attacks or spread as droplets to be carried by wind currents. Compounding its devastating effects, the oil sticks to clothing and is readily absorbed by the skin. Once mustard gas attacks became common (just as chlorine gas attacks before), medics would carry portable shower units with their attachments to quickly wash the mustard gas residue off the skin of soldiers in their care to prevent blistering.

Mustard gas posed an additional new threat on the battlefield. Mustard gas "lingered" long after the initial attack, as the oil is heavier than air or water, whereas chlorine gas did not settle but eventually dispersed from the battlefield as it permeated the air. This lingering effect led to long periods of contamination for battlefields and problems for soldiers on both sides, as exposure to leftover mustard gas was just as dangerous as the initial attack. Although mustard gas disabled soldiers for a shorter period of time (forty-six days compared to chlorine's sixty-day average), the lingering aspect made for a far more devastating and controllable weapon, as the users were not at the mercy of wind currents to guide it in the direction of opposing forces as with chlorine.[13] Germany would not be the sole country to use mustard gas, with France and England soon developing and using their own stockpiles.[14] The United States is not without fault either, as the country made use of chemical weapons as soon as it entered World War I. Future president of the United States Harry S. Truman oversaw the shelling of Germans with poison gas in 1918 while a captain of a U.S. field artillery unit.[15]

By the end of World War I, use of chlorine gas, mustard gas, and other chemical weapons accounted for more than 1.3 million casualties and more than 90,000 deaths.[16] The 1925 Geneva Protocol banned chemical weapons, condemning the use of "asphyxiating, poisonous, or other gases, and all analogous liquids, materials, or devices." Unfortunately for humankind and millions of soldiers, the Geneva Protocol had no real teeth to enforce such a ban, as it did not prevent stockpiling of chemical weapons or scientific research to create new ones.[17] This would not matter much to the United States, however, as the country only signed the protocol, failing to have Congress ratify the Geneva Protocol until fifty years later in 1975.[18]

A Military Cover-Up Leads to a Cancer Therapy

On December 2, 1943, a pivotal attack carried out by the Germans on an Allied forces port in Bari, Italy, led to the accidental discovery of mustard gas's anticancer effects on humans. During the attack, seventeen ships were destroyed, including the SS *John Harvey*, a ship that held within its bowels a clandestine cargo—2,000 mustard gas bombs. The oily liquid contents of the bombs spilled into the harbor and mixed with fuel from the destroyed ships, coating the survivors in a disgusting and deadly mixture that would leave them covered in blisters by day's end.

But if the use of chemical weapons was banned by the 1925 Geneva Protocol, why did the United States have 2,000 mustard gas bombs aboard the SS *John Harvey*? The 2,000 bombs were present just in case Adolf Hitler and the German forces decided to make use of their own chemical weapons, at which time the United States would retaliate in kind.

In addition to the conventional weapons used in the attack, which killed more than 1,000 American and British soldiers and hundreds of civilians, the detonation of the mustard gas bombs led to the vaporization of a portion of the oil contained in the SS *John Harvey*. This detonation created a toxic and deadly cloud of mustard gas that hung over the city of Bari for hours. The cloud exposed the 250,000 people living there to mustard gas and likely led to the deaths of up to an additional 1,000 people.[19]

The U.S. and the Allied forces refused to acknowledge the contents of the SS *John Harvey* in the moments after the attack, leaving medics and their patients unprepared for what the next days would bring. The reason for the cover-up? It's tied to the reason the mustard gas was present in the first place. The Allies feared that widespread knowledge of their holding mustard gas in reserve would give Hitler an excuse to tip his hand and go forward with the use of his own chemical weapons. The cover-up existed to prevent the escalation of tactics in the war but to the detriment of hundreds of lives. Soldiers who arrived onshore in many cases swam through the mustard gas–contaminated waters, coating their bodies in the blistering oil, and then sat with their wet uniforms still clinging to their bodies for hours on end. The telltale symptoms of blisters and swelling eyes began to show within a day. Physicians suspected chemical irritants were at play, but decades had passed since the world had seen the wide-scale use of mustard gas in war. Six hundred and seventeen soldiers were exposed, with eighty-three dying from the effects of wading and then sitting for hours in mustard gas. The physicians present characterized the status of the wounded as "dermatitis N.Y.D." (N.Y.D. for "not yet determined"), and British military officials—still kept in the dark over of the contents of the SS *John Harvey*—asked Allied headquarters for help in determining the source of the affliction. American physician and Lieutenant Colonel Stewart Alexander was soon sent in to aid and investigate. Alexander immediately observed the symptoms of the soldiers and diagnosed them with mustard gas exposure. The question at hand now was not what did the harm but from where the mustard gas originated. Alexander was aware of unverified rumors that the Germans had used chemical weapons in the Bari attack, but he did not immediately settle in with these accusations. Instead, he deftly reverse-engineered the air raid on the port at Bari using the available medical charts of the exposed and knowledge of their position with respect to the ships bombarded in the attack. This led Alexander to the determination that the SS *John Harvey* was the flashpoint of the disaster. Divers soon retrieved fragments of mustard gas shells from the wreckage and proved Alexander to be right. Alexander completed his summary of the incident, the "Final Report of the Bari Mustard Casualties," but the report was classified as soon as he turned it in. Supreme Commander of the Allied Expeditionary Force Dwight D. Eisenhower accepted the report as fact. His counterpart Winston Churchill, however, refused to believe it and to level any condemnation accompanying the reality that came with mustard gas being stored on a ship lying in wait for use. With the report classified, the medical records were also censored. Somehow, results from autopsies performed on the eighty-three soldiers who died escaped censorship, and they told a startling tale. All eighty-three of the deceased showed signs that the growth and multiplication of their white blood cells had been inhibited. This towering evidence suggested the mustard gas was the culprit and turned eyes and minds to the possibility that mustard gas could be used to slow and stop malignant white blood cells from invading other tissues in the body. In doing so, it would allow the gas to play a targeted role in fighting cancer as the first form of chemotherapy.[20]

This, however, is a fanciful story that history would have you believe. A startling discovery made from a pivotal defeat amidst a cover-up and clandestine investigation makes for a great tale after all. While mustard gas was aboard the SS *John Harvey* and the entire story is true, the actual discovery of the cancer-fighting effects of mustard gas—or, more specifically, a cousin of the sulfur mustards aboard the SS *John Harvey*—had taken place a year earlier in 1942.

Yale University, along with a number of other American universities, entered into contracts with the U.S. Office of Scientific Research and Development in the years surrounding World War II to perform experiments tangential to chemical weapons. Two Yale scientists with an interest in pharmaceuticals, Alfred Gilman and Louis S. Goodman, were asked to find an antidote to mustard gas and a series of its derivatives, particularly a new form of sulfur mustards—the nitrogen mustards. By this time, nearly 100 derivatives of sulfur mustard had been synthesized, giving plenty of fodder for wartime research.[21] Such would be the perfect task for Gilman and Goodman, as they were fresh off of producing the first edition of their now classic pharmaceutical textbook, *The Pharmacological Basis of Therapeutics*, which has since been revised numerous times and stands as a formative tome for pharmacists across the world. One catch came with the research contract (as often does when funds are dangled in front of academic researchers)—Goodman and Gilman were barred from publishing any data on their work until 1946 and the end of World War II.

Prior to the research push by the Office of Scientific Research and Development, we knew mustard gas and its analogues could play a role in life's drama as something other than a blister agent covering the fields of war. As early as 1934, it was known that mustard gas had some effect on white blood cells, with a 1935 study showing mustard gas to slow the growth of some tumors in animal models.[22]

Three different molecules, code-named HN1, HN2, and HN3, make up the nitrogen mustards, with only HN2 and HN3 designed for military use. HN1 was created to be a wart remover. Although they are designated HN1, HN2, and HN3, it is important to note that this is not the chemical shorthand for their structure. HN2 is the key molecule for our study, as it is the one out of the trio used in chemotherapy. Also, it is opportune to point out that the nitrogen mustards, unlike sulfur mustards, do not smell like mustard at all (or all that foul for that matter). The name "mustard" just stuck because of the structural similarity between nitrogen mustards and sulfur mustards, as both feature a nitrogen or sulfur core with chains emanating off of them that end in chlorine. The nitrogen mustards—particularly HN2, or chlormethine as it is known when applied to cancer treatment—are rather fruity smelling when concentrated, with this smell becoming either soapy or "fishy" depending on the individual's senses when chlormethine is diluted. Unlike sulfur mustards, the nitrogen mustards have never actually seen use in warfare. The nitrogen mustards are also not gases but rather, akin to the sulfur mustards from which they are derived, an amber-colored oil at room temperature.[23]

Our knowledge of tumors and cancer extends back millennia, with Hippocrates lending the umbrella of diseases its name. The Greek forefather of modern medicine named what we call tumors "karkinoma"—a cognate of the English word "carcinoma"—with Hippocrates gazing on a tumor and its various extensions and seeing the body and the legs of a crab.[24] Records of cancer predate Hippocrates, with the Edwin Smith Papyrus, an ancient Egyptian medical text named for one of its eventual owners and dating to 1600 BCE, telling of an individual for whom no treatment could be found and who exhibited "bulging tumors on the breast."[25] Through the early twentieth century, there existed only two ways to get rid of a cancerous tumor. The first—the millennia-old method—was to surgically remove the tumor from the body. The second was born out of William Roentgen's discovery of x-rays in 1895. In it, tumors are bombarded with x-rays in hopes of shrinking them in what was first called roentgen therapy for its discoverer but is now ubiquitously known as radiation therapy.[26] The aforementioned Goodman and Gilman would add a third—chemotherapy.

Goodman and Gilman did their first studies with nitrogen mustard on tumors present in mice before moving on to a larger animal subject, rabbits. In these studies, the pair observed nitrogen mustard to work to stop the growth of cancers involving lymphocytes. These animal studies were quite important, as they yielded evidence that nitrogen mustard could perform similarly on lymphomas in humans. Shortly thereafter in August 1942, one patient's dire condition and lack of options gave Goodman and Gilman a chance to try their new possible cancer therapy.

A forty-eight-year-old man known to history only as J. D. would stand as the first recipient of chemotherapy. J. D. received treatment with Goodman and Gilman's nitrogen mustard only after radiation therapy yielded no further positive results in combating his terminal lymphosarcoma. We do know a little about J. D. even if we do not know his full name. J. D. was a nondescript man who immigrated from Poland at the age of eighteen and spent most of his life working in a ball bearing factory in Connecticut. J. D.'s diagnosis of lymphosarcoma was first recognized in 1940 due to an enlargement and discomfort in his tonsils, with the cancer soon spreading to become tumors covering the right side of his neck and leaving him barely able to open his mouth. A painful two-year trek of radiation therapy cycles and surgery began, but after two years, J. D.'s cancer resurfaced once more, leaving him and his physicians desperate for another option.[27]

On August 27, 1942, Goodman and Gilman, at the behest of Gustaf Lindskog, J. D.'s physician and an assistant professor of surgery at Yale, began administering intravenous doses of nitrogen mustard to J. D., starting small with a tenth of a milligram of the oil per kilogram of body weight. Eventually, they worked their way up to a whole milligram per every kilogram of the patient's weight (what we now know to be nearly three times the standard dose). Within two days, there was a noticeable softening of the tumors of J. D.'s lymphosarcoma. Further symptoms improved after the fifth treatment, and by the end of ten daily treatments of nitrogen mustard, a biopsy would show the tumors to be altogether gone.[28] This was a great success, as chemotherapy with nitrogen mustard was shown to be a viable option in the fight against cancer, joining surgery and radiation therapy as a third arrow in the physician's quiver. Unfortunately for J. D., his cancer returned inexplicably forty-nine days after the initial intravenous injection of nitrogen mustard, with further treatment of nitrogen mustards proving less effective than the first round. J. D. would die from this relapse on December 1, 1942, but Goodman and Gilman, along with the physicians involved, showed resilience, pressing forward with a small clinical trial.[29] The trial, shrouded in secrecy, took place with sixty-seven participants, with no information given to their family or caregivers about their involvement in a study with nitrogen mustards.[30] On the backs of Goodman and Gilman's work, chlormethine became approved by the U.S. Food and Drug Administration in 1949 for the treatment of non-Hodgkin's lymphoma. It was sold under the trade name Mustargen, making it the first form of chemotherapy to reach the mass market.

Despite protocols, a general sense of uneasiness surrounding the use of chemical weapons, and the eventual application of nitrogen mustards to cancer, the use of sulfur mustards as a weapon of war and the problem of prolonged contamination never really went away. The bulk of the unused mustard gas found in Germany after World War II did not see proper disposal, making for an ecological nightmare as the canisters were dumped into the Baltic Sea. In the passing years, the mustard gas forms a polymer that looks like amber, and from time to time, these amber-like nodules wash ashore and pose a dangerous health hazard.[31] Sulfur mustards were used as recently as the 1980s by Iraq in attacks on Iranian soldiers during the eight-year Iran–Iraq War, with as many as 350 separate chemical weapon strikes carried out.[32] Iraq also used mustard gas in shameful attacks on Kurdish citizens, which drew the ire of the world community. These attacks, now known as the Halabja Massacre, took place on March 16, 1988. President of Iraq Saddam Hussein ordered the attacks, which killed between 3,000 and 5,000 innocents during the shelling of the Kurdish village of Halabja with a combination of mustard gas and nerve agents.[33]

How Nitrogen Mustards Fight Tumors

Some derivatives of HN2 still see use as chemotherapy agents today, where it is often used as part of a combination therapy to treat patients with both Hodgkin's and non-Hodgkin's lymphomas, lung cancer, and leukemia as well as use in a topical form to treat mycosis fungoides.[34] Chlormethine is normally given intravenously, with any spillage at the injection site on the vein likely leading to blisters on the surrounding skin. Unfortunately, as with most chemotherapy drugs, side effects follow soon after administration, with patients often vomiting within eight hours of receiving a dose of chlormethine.[35] Treatment of mycosis fungoides takes place differently, as the cancer displays as a series of plaques or patches on the skin. Prior to the 1980s, treatment for mycosis fungoides was given as an aqueous solution of nitrogen mustard, with this eventually changing to an ointment-based preparation that was much better received by patients.[36]

The mechanism by which the 156-dalton nitrogen mustard chlormethine acts on the body is quite intriguing. The molecule reacts with itself, forming a highly reactive three-part ring structure and releasing a chloride ion in the process. It has long been thought that the release of the chloride ion is behind the blister effects seen with mustard gas, as they also undergo this same ring formation and chloride release. Why? It would be easy for the chloride ion to bind a hydrogen from the surrounding area—say, a molecule of water—and form hydrochloric acid. This is not where the nitrogen mustard molecule stops, however, as the possible formation of hydrochloric acid is an action separate from the molecule's cancer-fighting abilities. Once the highly reactive three-member ring structure is created, it acts as an alkylating agent, easily binding with DNA in the cells. The nitrogen mustard ring structure seeks out guanine residues in DNA, one of the parts of our genetic code, causing the guanine to bind a different partner than normal. This leads to guanine exchanging its typical companion cytosine for thymine and opening up the possibility of mutations in a strand of DNA during the subsequent replication events or, at the very least, stressing the cell by creating additional points needing repair. It is also possible that the binding of the nitrogen mustard ring structure to guanine leads to a very detrimental reaction, the cross-linking of two strands of DNA, which often causes a biological "dead end" for the strands of DNA involved. The damaged DNA is then spotted by the cell, which then signals tumor protein 53 to start the process of apoptosis, in which the cell essentially commits suicide due to the presence of extensively damaged DNA.[37] What the use of nitrogen mustards in chemotherapy aims to do is cause enough damage to DNA in the right spots—tumor cells—so that the cells start the cycle of apoptosis and in doing so shrink the tumor. I am not sure one would have predicted that a successor to the sulfur mustards that drenched World War I battlefields would one day be used to shrink tumors. Thankfully it is, with chlormethine behind the treatment of a legion of patients in the intervening decades despite its horrifying origins.

CHAPTER 10

Warfarin

The structure of warfarin, a rat killer turned lifesaver. *Structure generated by author in ChemDoodle v 11.5.0, using information supplied by the National Institutes of Health*

Warfarin is one of the most commonly prescribed drugs in the world, with 1 percent of the population in the United Kingdom and more than 15 million in the United States taking the anticoagulant drug on a daily basis to prevent blood clots and their complications. Warfarin, however, started out as something very different—a rat poison. How did a rat poison become one of the most commonly prescribed drugs in the world? The story involves a legion of dead cows, an erratic professor, and an eerily named chemical, analogue #42.

Dead Cows and Detective Work

Anticoagulants—drugs that thin the blood—have long been a part of modern medicine, but they do have historic analogues as well. In the fifth century BCE, Hippocrates believed that thinning the blood would aid in the treatment of a number of disease states but lacked a systematic way to do so. Leeches were applied to achieve this purpose, but the early twentieth century would provide us with a better method thanks to heparin and warfarin. Heparin arrived on the scene in 1916. Heparin is an extraction from animal liver and intestines, but it is unable to be delivered orally, increasing the difficulty of administration, as it must be injected.[1] The search for an oral anticoagulant was on for the next several years, but it would take a bewildering series of events culminating in the deaths of thousands of cows across North America to obtain one.

In the late 1920s across the northern United States and Canada, bizarre stories of cows bleeding profusely spread like internet tales of crop circles and cattle mutilation. Cows were dying in the process of dehorning or castration or just dying peacefully from seemingly no cause at all, only to be determined that they were suffering from internal bleeding postmortem. The high value of cattle made this phenomenon quite the hardship for farmers and ranchers, one already compounding the high costs necessary to keep their livestock fed.

To solve this mystery, Canadian veterinarian Frank Schofield became one part detective, eventually deducing that the cows were dying not because of a new virus or bacterial infection spreading across the continent but because of what they ate. The cows afflicted had one thing in common—they were eating old, moldy, sweet clover silage. Silage is sweet clover hay that is stored but not dried in a silo and that is often put away as a nutrient-rich food source in the winter months. The silage that cows across the country were dining on contained mold, which in many cases could be clearly seen on the hay. But if the silage was moldy, why were farmers continuing to feed it to their prized cattle? Financial hardships weighed heavily on farmers in the 1920s, and, moldy or not, food was food. Farmers simply could not afford to purchase "fresh" silage once their remaining supplies became moldy, and in doing so, they accidentally poisoned their livestock.[2] To test whether the moldy hay was at fault, Schofield fed samples of moldy and fresh silage to rabbits, with those unfortunate rabbits assigned moldy silage experiencing the same fate as the cows. The hay was to blame, and the only options for farmers was to discard their spoiled silage for fresh silage or perform a blood transfusion. Roughly a decade later, a second veterinarian, L. M. Roderick, showed that blood's prothrombin supplies were affected by the moldy silage and that this interaction was leading to the clotting deficiencies.

Karl Paul Link and a Can Full of Blood

Karl Paul Link, who would eventually aid in pioneering warfarin, did not experience the most picturesque of childhoods. Labeled as weak from an early age, pneumonia nearly claimed his life at the age of two. His family lived in a near state of poverty, with most of his clothes hand-me-downs. Intellectually, the Links were rich thanks to the gifts of his mother and father, Frederika and George. Both German and English were spoken in their Indiana home, with his father a practicing Lutheran minister until a throat malady forced him to leave his ministry. Afterward, his father took on several different jobs en route to becoming a lawyer. Unfortunately, as we see in the background of several of the scientists who made major contributions to the field of medicine, Karl Link's father died of cancer at a relatively early age when Karl was twelve. Despite the loss of the primary breadwinner, Frederika Link's strength kept food on the table and all ten of her brood's minds full. All of Frederika's children would go on to become rather successful, counting a registered nurse, petroleum geologist, politician, lawyer, and Red Cross administrator among their number aside from Karl Paul Link.[3]

Link's liberal ideals showed through at a young age, as Wisconsin governor (and later senator) Robert M. La Follette's Progressive movement is considered to have played a role in Link making the trek from Indiana to Wisconsin for his college years. Link began at the University of Wisconsin in 1918 as a freshman, leaving seven years later with his PhD. He supplemented his education overseas with a postdoctoral stint in the lab of Sir James Irvine at the University of St. Andrews in Scotland. During his tenure in Scotland, we get the first sense of the air of conflict that would linger over Link throughout his career. Link was kicked out of Irvine's lab within a year. He eventually landed in a better situation in the lab of 1923 Nobel Prize winner Professor Fritz Pregl at the University of Graz in Austria. Pregl concerned himself with the matters of microchemistry, a near art that seeks to learn as much about a chemical compound as possible while using as little of the compound as necessary. As far as we know, Link and Pregl got along swimmingly, with Link bringing back one of Pregl's microchemistry equipment sets to the United States and showing it off gleefully to anyone who showed interest. During his time in Europe, it is thought that Link developed his trademark flamboyance in dress, which often included wearing a cape. A cape-wearing Link would certainly stand out on returning to the midwestern United States and the University of Wisconsin in 1927, where he worked as an assistant professor in agricultural chemistry. At the University of Wisconsin, Link centered his academic world on plant biochemistry, particularly the study of carbohydrates, sugars and chains thereof often used as an energy source.[4] Throughout his career, the ever-intense Link was uniquely devoted to his work.

A story claims he went back to his lab to work in the afternoon after his morning wedding to Elizabeth Feldman in 1930. Thankfully, Link's tale of wedding day devotion to his laboratory and not his spouse is likely a work of fiction.

Entrenched in studying plant biochemistry, Link was propelled into the world of anticoagulants by sheer chance. Between December 1932 and February 1933, Ed Carlson experienced the death of five cows and was about to lose a sixth, sending the farmer into a spiral looking for help. Carlson did not trust the local veterinarians, as he fed his cows sweet clover hay for years without issue. Sullied by the deaths in his flock, Ed Carlson—in the middle of a blizzard and temperatures of −18°C—set out from his farm in Deer Park, Wisconsin, to make the nearly 200-mile trek to the state capital of Madison. Why? He had hopes that veterinary scientists in the Wisconsin state office could solve his mysterious problem. Making the trip with Carlson in the bed of his truck was one of his recently deceased female cows, a milk can full of blood (blood that refused to clot), and a sample of the sweet clover hay—as much as 100 pounds by some accounts—he was feeding his cows.[5] Unfortunately for Ed, the state office was closed, as it was a Saturday, leaving the desperate Carlson to knock on the doors of anyone available and willing to listen. Carlson knocked until he found Karl Paul Link, who was whittling away at work on a Saturday in the inclement weather along with a few of his students, foremost to this tale being Wilhelm Schoeffel. Link, interestingly enough, had become aware of the problems surrounding sweet clover and its molds only a month earlier, when a colleague in the genetics department at the University of Wisconsin pulled Link in on studies on the amount of the molecule coumarin in sweet clover hay and its palatability to livestock. Coumarin is the chemical responsible for the wonderful smell of freshly cut hay that flows through the air in the fall, but too much coumarin in the hay makes the foodstuff bitter. With Link assisting, the geneticists thought it might be possible to devise a strain of sweet clover hay with a lower coumarin content and in doing so make an even sweeter strain that would please cows and grow well in the climate of Wisconsin.[6]

Link, drawing on his newfound knowledge of sweet clover hay, imparted the only wisdom he could, taking from the previous work of Roderick and Schofield in the 1920s. He advised Carlson to cease feeding the cows moldy hay and, if possible, perform a blood transfusion on cows exhibiting symptoms. One of Link's students in the lab that day, Wilhelm Schoeffel, famously summed up the plight of Ed Carlson in his thick Schwabian accent, saying, "Vat de hell, a farmer shtruggles nearly 200 miles in dis Sau-wetter, driven by a shpectre and den has to go home vit promises dat might come in five, ten, fifteen years; maybe never, who knows? 'Get some good hay—transfuse.' Ach! Gott, how can you do dat ven you haf no money?"[7] Schoeffel deeply empathized with the farmer and at the same time was fascinated with the milk can of unclotted blood. He immediately canceled his weekend plans to work on the problem in hopes of finding the exact chemical compound in spoiled sweet clover hay preventing the blood from clotting.

Six years later, Link and his team would discover that it was the same substance responsible for the flavor of sweet clover hay, coumarin, that was also responsible for the clotting issues observed in cows across North America.[8] In moldy hay, coumarin underwent a chemical change wherein molecules of coumarin were oxidized, which allowed for the binding of two coumarin molecules together in the presence of formaldehyde to create dicumarol, the end molecule responsible for the blood-clotting problems.[9]

Once Link determined dicumarol to be the culprit, his laboratory synthesized more than 100 analogous molecules. Of these, analogue #42 will eventually become the star of this story, but we will get to its role later. Link did not discover dicumarol, as it had been synthesized four decades earlier in 1903 by a group of German chemists; however, it was entirely unknown at the time that dicumarol existed in nature. The University of Wisconsin and Link, in conjunction with the Wisconsin General Hospital and the Mayo Clinic, began a small clinical trial to test the effectiveness of dicumarol in humans, tests conducted in hopes the molecule would reduce the formation of dangerous blood clots.[10] These tests were successful, with the Wisconsin Alumni Research Foundation (WARF) awarded a patent on dicumarol in 1941.[11] These results were soon followed by the discovery that vitamin K acted as a natural antidote to overdosage of dicumarol, aiding dicumarol's quick acceptance by the medical community.[12]

Tuberculosis and Tempers Flare

In September 1945, Karl Paul Link experienced a recurrence of tuberculosis, a disease that interjected itself throughout his life. This instance of tuberculosis reared its head when Link was away on a canoeing trip with his family, a trip Link was taking to stem exhaustion from overwork. The case was bad enough for Link to be moved from the Wisconsin General Hospital and sent to the Lake View Sanitorium for a six-month stay, where the relaxed mood, fresh air, and cod liver oil supplements were no doubt the epitome of trappings that the hardworking and obsessive Link would despise. Karl did find outlets outside of the three bottles of beer he was allowed each day, however. During his time in the sanitarium, Link became consumed with reading about methods of rodent population control throughout history.[13] Link previously visited the possibility of rodent control using dicumarol with his favorite student, the formerly mentioned Wilhelm Schoeffel, but the chemical would never fit the bill when it came to killing rats. The anticoagulant could cause bleeding, yes, but it acted too slowly to ever be an efficient killer.[14] The diet of a rat is also high in vitamin K due to the variety of grains and leafy green foods they typically consume, providing a natural antidote to the slow trod of dicumarol poisoning.[15]

Although their leader would be sidelined for nearly a year, work in his laboratory did not stop, as Link invited Mark Stahmann to keep things going in his absence. Here, we return to the previously mentioned synthesized analogue of dicumarol, analogue #42. Stahmann seized on the molecule, as the group knew it was considerably more potent than dicumarol and as such would work faster in killing rodent populations. Link never thought much of analogue #42, accepting its potency as its failing and going so far as to never even patent the molecule. With twenty-three days left before the possibility of a patent would vanish, Stahmann spearheaded an effort with WARF to patent analogue #42.[16]

After recovering from tuberculosis, Link ventured back into his lab. Shortly after Link's return, one of the first public signs of his intense temper arose with Stahmann as his target. As the duo disagreed on an unknown subject, Link argued intensely with Stahmann, directing him to cease all work on anticoagulants. Link then kicked him out of the laboratory, definitively ending their collaboration.[17] This animosity extended across decades, as a 1959 article written by Link about the discovery of warfarin leaves out any input on the part of Stahmann.[18]

His altercation with Stahmann is not the only time Link found himself losing his cool with a colleague. A close friend of Link's, Robert H. Burris, recounted such an incident in a biographical sketch of his companion and mentor. In this occurrence, Link jealously "throttled" Harry Steenbock, a pioneer at the University of Wisconsin in vitamin D studies, in the men's bathroom and continued to berate Steenbock in the hallway as a small group looked on.[19]

Despite Link's unfortunate tendency toward jealousy and outbursts of anger, he was a noted liberal and outspoken advocate for the rights of students. This also got him in trouble, as at one point in the early 1950s he was censured by the University of Wisconsin Board of Regents for publicly speaking out against their policies in the press. Link commonly took the side of students in cases against the administration, even going so far as to establish a legal defense fund explicitly for students who found themselves on the wrong side of the law or embroiled in a battle with the University of Wisconsin. As did many intellectuals of the time period, Link opposed the efforts of Senator Joseph McCarthy. Link went a step farther, sponsoring liberal and socialist student groups in an effort to take a stand, including the John Cookson Karl Marx Discussion Group and the Labor Youth League. As the decades passed, however, Link felt less comfortable with new colleagues often unfamiliar with his past and strict ideology and finally refused to come to department meetings after a newer member of the staff labeled him offhandedly as a conservative. Despite his own efforts, Link's lasting legacy of liberal ideology came through a gesture of his wife's after his own passing, with Mrs. Elizabeth Feldman Link bequeathing their family home to a local peace group after she died.[20]

Analogue #42 and a Fateful Suicide Attempt

Analogue #42 would prove to be quite the efficient killer, with the chemical compound dubbed warfarin—"warf" for WARF, which sponsored it, and "-arin" as a throwback to its biological precursor coumarin.[21] Warfarin, although more potent than dicumarol, still acted slowly compared to other rodent control methods like strychnine, but rats persisted in visiting the warfarin-laced bait time and time again as they inadvertently poisoned themselves. News of the discovery of a better rat killer circulated quickly, with Chicago-area businessman Lee Leonard Ratner licensing warfarin from WARF in September 1950 for use as a rodenticide. This resulted in d-CON, short for "decontaminate," and the product is still sold today.[22] At the time, d-CON was sold in four-ounce cans, intended to be mixed with grain or ground beef to a total of six pounds. To show off his new product, the aptly named Ratner thrust himself forward as the benefactor of the city of Middleton, Wisconsin, which, quite unusually, was a city known at the time to have a particularly bad rat problem. On November 4, 1950, Ratner began a fifteen-day campaign in conjunction with the city's rodent control committee and local Boy Scout troops to set out bait stations containing warfarin. Just over two weeks later, there was little sign of rats in the city, with Ratner, ever the salesman, suggesting the citizens of Middleton set out fresh d-CON in bait stations at regular intervals to prevent the rats from revisiting.[23]

With warfarin firmly established as a commercial rodenticide, Link completed his side quest of applying a coumarin analogue to rodent control. But that does not get us to where we need to be—the application of warfarin to human health issues, particularly the prevention of blood clots. So who decided to put a potent rat poison into people? It was not a physician or a scientist, at first at least, but a suicidal Korean War–era U.S. Navy recruit. In 1951, a twenty-one-year-old recruit reported to the Philadelphia Naval Hospital, complaining of back and stomach pain as well as the inability to walk. The patient, known only as EJH, did not speak at first, but once he was able to, the physicians learned his story. In order to avoid any sort of impending military deployment, EJH began consuming the rat poison warfarin. Shocked it did not kill him the first night he consumed the poison, he ate more the next two nights. Eventually, bleeding crept in, starting with his nose, as well as the stomach pain that would eventually lead him to the hospital. In total, EJH consumed a package of rat poison containing 567 milligrams of warfarin mixed in cornstarch, but the fate he clamored for eluded him.[24] Instead, the pain grew so great that he went to the Philadelphia Naval Hospital, where he was administered vitamin K as an antidote as well as blood transfusions, which reversed his self-inflicted situation and pain.[25] This bizarre set of circumstances convinced Link and physicians to try warfarin in people to prevent blood clots, seeing the molecule as a more potent version of dicumarol and possibly its successor.

Warfarin was rebranded under the trade name coumadin to separate it from the rat poison and was approved for use by the U.S. Food and Drug Administration in 1954.[26] Coumadin (and thus warfarin) weighs in at 308 daltons, once again among the drugs in this book that follow modern drug design size limitations. On September 29, 1955, Karl Paul Link received a card in the mail from an individual working in the Fitzsimmons Army Hospital in Denver, Colorado, the hospital caring for President Dwight D. Eisenhower after he endured a heart attack at his in-laws' days earlier. (This is where I suggest a new rule: you don't have to go to your in-laws anymore if you are sixty-four years old and the president.) What did the card, which was sent by someone who previously lived in Wisconsin and familiar with Link's work, say? "The President is getting one of your drugs and it's not dicumarol."[27] The process of elimination would suggest Eisenhower was receiving warfarin turned coumadin, with his press secretary later backing up this assumption. Despite its connections to a powerful rat poison, the use of coumadin was bolstered in the public eye by Eisenhower's being given the medication following his 1955 heart attack.

Coumadin's Use Today

It would be two more decades before scientists would discover why coumadin works as an anticoagulant. In the late 1970s, John W. Suttie determined that coumadin binds vitamin K epoxide reductase (also known as VKORC1). VKORC1 plays a role in modifying vitamin K in the body en route to vitamin K preparing blood clotting proteins for their role in the body. If molecules of vitamin K are never modified by VKORC1, the blood-clotting proteins are created in lower quantities and with less effectiveness, decreasing the body's overall ability to clot blood when an incident occurs.[28]

Today, coumadin is used as a maintenance medication—that is, a drug taken to ease a chronic, long-term disease—aimed primarily at preventing strokes due to blood clots in patients with atrial fibrillation (which leads to an irregular and often rapid heart rate), those with diseased heart valves, or those with prosthetic heart valves. Coumadin also sees use to prevent and treat deep-vein thrombosis, a situation where clots of blood form, often in the legs. If the clot moves, it can migrate and lead to a blood clot in the lungs, a life-threatening situation known as a pulmonary embolism.[29]

As with many of the drugs in this book, coumadin exhibits a narrow therapeutic index. The success of coumadin dosage is measured not just in a lack of blood clots but also in a lack of bleeding, as too much coumadin can cause serious bleeding events. Medical personnel measure a patient's international normalized ratio (INR) to gain insight as to how well a patient is being dosed or if the patient is properly complying with dosage instructions. The INR is calculated based on how long it takes a patient's blood to clot versus the clotting time in a normal individual. The target INR is between two and three, with problems occurring when a patient's INR drifts below two—an increased risk of blood clots forming—or rises above three—a higher likelihood of bleeding.[30]

Due to the narrow therapeutic window of coumadin, keeping tabs on a patient's INR and general well-being is of utmost concern for health care providers. This has led to the birth of anticoagulation clinics in many parts of the world, where health care providers are specifically trained and tasked with keeping patients within the proper INR window. Studies in the United States, the United Kingdom, and China have shown that clinical pharmacists—pharmacists acting as one part physician and one part pharmacist and doing so often in a hospital setting—are more cost effective within these clinics at achieving target INR values than a corresponding physician due to a pharmacist's ability to spend more time with the patient and keep tabs on said patient by monitoring between appointments.[31] Patients can also self-monitor their INR, with a combination of these approaches often wielding the best results. Coumadin dosages often start at five milligrams, with the provider increasing the dosage until the target INR is achieved. This is difficult, as the dosage needed for each person can vary wildly. Genetics—particularly the variations in the genes coding for CYP2C9 and VKORC1—plays a large role in how warfarin is metabolized and thus how much is needed. Age and body surface area also play roles and account for an estimated 55 percent of the variations observed in dosages between patients.[32] These elements are why variations are so large and why those taking coumadin as a maintenance medication require such close monitoring. Despite INR monitoring, fatal bleeds still occur, much like they did in cattle in the 1920s. These fatal events occur at a rate of one per seventy-six years of taking the drug, meaning that if eight people take coumadin as a maintenance medication for ten years each, it is estimated that one of them will suffer a fatal bleeding event.[33]

When taking warfarin, one must avoid certain foods that contain large amounts of vitamin K. Why? Vitamin K, the natural antidote to coumadin, aids in making your blood clot, sabotaging the effects of warfarin. Partaking of typically super-healthy leafy green vegetables like kale, spinach, and collard greens can be particularly dangerous to those on a coumadin regimen. This trio of vegetables contains nearly 1,000 micrograms of vitamin K per a single-cup serving.[34]

Warfarin, Stalin, and Vampire Bats

Just as warfarin is able to poison rats, it can certainly be used to poison humans too. Speculation exists as to whether Joseph Stalin's fatal brain hemorrhage was the result of warfarin poisoning. Several days before his death on March 5, 1953, Stalin dined with four of the highest-ranking members of the Politburo, including Lavrenti Beria, the minister of internal affairs, and Stalin's eventual successor, Nikita Khrushchev. Why would Stalin's associates murder him? Possibly due to the "doctors' plot," a plan to pin the assassinations of several high-ranking officials on poisonings committed by Jewish physicians and possibly due to fears Stalin wanted to escalate the current tensions with the United States to a very palpable level. How high of a level? Researchers suggest Stalin was about to announce the United States was planning an attack on Moscow in order to justify a Soviet attack on the Pacific Coast of the United States as well as Soviet expansion into China. Those aligning themselves with the theory of Stalin's poisoning claim such designs would be disastrous for the Soviet Union and likely lead to a nuclear showdown. Of the four, researchers pin Beria as the culprit of the plot, which in a way is believable, as Beria is rumored to have boasted of killing Stalin to save his fellow man on May Day 1953.[35] Beria died in the sights of a firing squad shortly thereafter in December 1953 for treason. The theory falls apart if one seeks to claim the poisoning happened over a single dinner. (As we previously learned with the poisoning of EJH, it takes the consumption of a significant amount of warfarin and multiple days for a poisoning to show signs, and even then, it might not lead to death.) The poisoning could be successful if it helped exacerbate an already existing condition, which is quite reasonable considering Stalin was seventy-four at the time of his death and not nearly as robust physically as a twenty-one-year-old naval recruit. While we may never know for sure whether warfarin played a role in Stalin's death, it is not outside the realm of possibility.

While warfarin could theoretically be implicated in the prevention of mutually assured destruction during the Cold War, it is directly implicated in the death of something else out of our nightmares—vampire bats. In parts of the world such as Latin America and South America, where vampire bats dine on livestock and often spread rabies in the process, the drug is mixed with petroleum jelly and used to coat large nets or is placed where the bats might perch. Why? Vampire bats clean themselves through the process of mutual grooming, with one bat licking and snacking on another bat until it is cleaned and vice versa. If the bats are coated with the warfarin-containing petroleum jelly, the bats consume the warfarin during mutual grooming and slowly spread the poison throughout the colony. This is problematic, as it can also kill other types of bats. More specific methods of controlling the vampire bat population using warfarin include injecting cattle with low levels of warfarin in hopes they will pass it on to the vampire bats during feeding or placing a paste containing warfarin over open vampire bat wounds on livestock, relying on the possibility of vampire bats feeding on the same spot to spread the anticoagulant.[36]

Despite the use of warfarin to control the population of vampire bats, it may be the rats—those that warfarin was first applied to—that are having the last laugh. Some rats are showing resistance to warfarin baits, with a study of European mice and rats showing mutations in the VKORC1 gene.[37] Through these mutations, the effect of warfarin wanes, allowing the rats to eat their fill and rest well through the night, possibly starting the countdown to the end of warfarin's usefulness in controlling the rat population.

CHAPTER 11

Botulinum Toxin

The structure of botulinum toxin serotype A, one of the primary serotypes used in medical practice. *Image created from information deposited in the RCSB PDB (rcsb. org, H. M. Berman et al., "The Protein Data Bank"). Image generated with PDB ID 3BTA using Mol* (D. Sehnal, S. Bittrich, M. Deshpande, R. Svobodová, K. Berka, V. Bazgier, S. Velankar, S. K. Burley, J. Koča, and A. S. Rose [2021])*

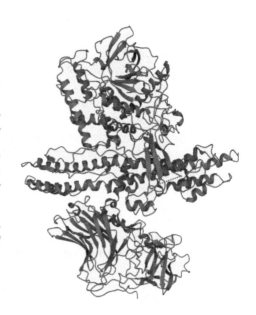

Botulinum toxin, or Botox as it is commonly called, is not a tablet or capsule like many of the drugs in this book but a delicate solution that must be injected by a medical professional. Botulinum toxin is a great example of one main drug receiving several U.S. Food and Drug Administration (FDA) approvals as more uses were discovered and shown to be safe and effective. This is in part because botulinum toxin is very different from the other medications we have looked at so far. It is not a small molecule but a 149,000-dalton protein that must be acted on by the body to perform. Seven different types of botulinum toxin exist—A, B, C, D, E, and F—with two of these, A and B, accounting for the majority of medical uses.

Botulinum toxin is also a neurotoxin made by bacteria, but that might be an understatement: it is possibly the most potent naturally occurring toxin known to humankind.

The bacteria responsible for botulinum toxin, *Clostridium botulinum*, is pretty much ubiquitous in the environment, from soil to lakes and streams. *Clostridium botulinum* is a gram-positive bacterium featuring a cell wall made of a thick layer of peptidoglycan. It is anaerobic, meaning it does not need oxygen to survive and would prefer to live in the absence of oxygen, wherein the bacteria can secrete the simply named neurotoxin botulinum toxin, a toxin leading to the illness botulism in humans.

Sausage and the Most Poisonous Toxin Known to Humankind

The earliest account of an outbreak of food-borne botulism occurred in 1817 in Württemberg, Germany, with physician and poet Justinus Kerner recognizing a certain set of symptoms in many individuals who had eaten smoked sausages. Kerner also noted that botulism, in extreme cases, paralyzed muscles and rendered many functions of the parasympathetic nervous system moot. This observation led Kerner to suggest that whatever was behind botulism might one day have a legitimate medical use. However, although he did discern that the poison was biological in origin, in his time, Kerner was unable to isolate the toxin behind botulism, instead calling the just-out-of-reach culprit the "sausage poison" or "fatty poison."[1] Between 1817 and 1822, Kerner wrote extensively about the sausage poison and its telltale signs. These signs include drooping eyelids, blurred or double vision, difficulty swallowing, and slurred speech, followed by the general muscle paralysis and difficulty breathing due to respiratory failure as the disease progresses to eventual death in many cases.[2]

While Kerner's notes are the earliest record we have of a food-borne case of botulism, the disease has likely been around as long as humans have consumed food. Science historians have suggested that an edict enacted against the creation of blood sausage made by Emperor Leo IV of Byzantium in the late 800s CE likely emanated from an outbreak of botulism.[3]

Although Kerner did not discern the toxin behind the disease botulism, the discovery of the source would come later in that same century. An 1895 outbreak of food poisoning stemming from smoked ham served at a funeral dinner in Ellezelles, Belgium, led Belgian biologist Émile van Ermengem to isolate *C. botulinum* from one of the putrid food sources and begin steps to truly investigate the reason behind the outbreak of food poisoning.[4] Van Ermengem lent *botulinum* to the species name, as the bacteria was closely linked to poisoning via sausage, with the Latin word for sausage being *botulus*.[5]

Although van Ermengem isolated *C. botulinum*, it is not infection with the bacterium that causes the disease state botulism. Instead, it is one of the molecules created by the bacteria botulinum toxin, which is responsible for both botulism and its recently discovered therapeutic effects. Botulinum toxin is the most potent poison known to humankind, with the toxin lethal at a dose as low as 0.05 micrograms.[6] To give an idea of how little material this is, an eyelash hair weighs in at about seventy micrograms, meaning an eyelash worth of botulinum toxin, if divided correctly and spread with the utmost of efficiency, would theoretically be able to end the lives of 1,400 people. A fatal dose of botulinum toxin looks like flaccid paralysis—a decrease in muscle tone or general paralysis in a region—followed by respiratory failure.

While seven different forms of botulinum toxin exist, three forms—C, D, and E—are responsible for the majority of the cases of botulism in mammals.[7] It is important to note that the two main serotypes of botulinum toxin used in medical practice—A and B—are missing from this list.

Clostridium botulinum secretes its toxin when in the proper mixture of temperature and low-oxygen environment. *Clostridium botulinum* is found in the soil and in rivers, allowing for many possibilities of contamination if cleanliness is not a priority. Unfortunately, errors in at-home canning of food often produce the exact blend of temperature and low-oxygen environment necessary for *C. botulinum* to flourish. When this happens, we see the toxin released into the improperly preserved food, just as with the sausage and smoked ham in the poisoning cases Kerner and van Ermengem observed. Once the toxin contaminates the food, a case of botulism is likely in the future for whoever consumes their loved one's canned food project. Botulinum toxin is rendered harmless through boiling, making ready-to-eat preserved foods (think sausage and canned tuna fish) the main sources of botulism risk. Botulism, thankfully, is not spread by person-to-person contact like influenza or COVID-19, decreasing the spread of the disease.[8]

The First Uses of Botulinum Toxin

Not everything about botulinum toxin is bad, however. Just as Kerner hypothesized, we have learned how to make many positive applications of the toxin. The initial therapeutic use of botulinum toxin came in patients exhibiting dystonia, a condition wherein the muscles of the body contract uncontrollably, leading to involuntary twisting and other movements of the body. In particular, botulinum toxin was applied to treat cervical dystonia, which is a contraction of the neck muscles often requiring surgery, and oromandibular dystonia, which is often diagnosed by uncontrolled opening and closing of the jaw along with the grinding of teeth. No surgical options existed for the latter, with a case of dystonia often debilitating and making for a considerable quality-of-life issue.[9]

Although use of botulinum toxin to treat dystonia was well known and har-kened back to the original observations of Justinus Kerner almost two centuries earlier, the use of botulinum toxin in and around the eyes, its primary modern use, would come in the late 1970s and 1980s. Alan Scott injected primates in the face and eyes with botulinum toxin during the 1970s, observing that the mam-mals were able to easily tolerate the injections without undue harm. With this knowledge in hand, Scott sought to apply botulinum toxin type A to solve two common problems in his practice: strabismus and blepharospasm. Strabismus is a condition that affects up to 4 percent of the population wherein the left and right eye do not look in the same direction at the same time. The FDA granted approval for investigational clinical trials of injections into the muscles around the eyes to treat strabismus in 1977. These injections provided a much-needed alternative to surgery, with a single initial injection able to correct an incidence of strabismus up to forty prism diopters, a phenomenal correction in deviation.[10] Scott's initial study involved forty-two patients receiving 132 doses of botulinum toxin, with a positive effect on strabismus seen over one year after injection in some patients with minimal side effects.[11]

The second problem Scott sought to solve with botulinum toxin, blepha-rospasm, would follow soon thereafter, with his first patient coming in 1980. Blepharospasm is an uncontrolled twitching of the eyelids, which sometimes manifests as eyelids that shut extremely tight. Scott recounted the sight of his first blepharospasm patient treated with botulinum toxin, saying that her eyes were closed so tightly that she needed her husband to guide her into his office. A single injection of botulinum toxin into the center of the eyelid successfully allowed her eyes to open, but they went back to their prior state a day later. This was the first sign that repeated injections of botulinum toxin would be neces-sary to cure some ailments, but Scott successfully helped the patient see again for the rest of her life by altering the injection site and repeating the procedure as needed.[12]

Alan Scott developed a process for efficiently growing and purifying botu-linum toxin type A from *C. botulinum*, eventually forming the company Oculi-num to protect his research. The biotech company Allergan became interested in Scott's work, and a deal was reached to acquire Oculinum from Scott in 1991.[13] Allergan manufactured botulinum toxin type A under the name of Scott's com-pany, Oculinum, for use to treat strabismus and blepharospasm, the previously mentioned treatments Alan Scott pioneered and gained FDA approval to treat in 1989. Interestingly, Oculinum received orphan drug status from the FDA as part of the 1983 Orphan Drug Act, which seeks to offset the research and devel-opment costs of drugs that will likely have a small user base.[14] Such a designation granted Allergan an exclusive marketing period for the drug that would soon see its user base grow exponentially as new applications were found. In time, Allergan would market botulinum toxin type A under the name Botox, the name by which we now know the drug.

The First Application of Botulinum Toxin in the Cosmetics Industry

In 1987, plastic surgeon Richard Clark accidentally severed the frontalis branch of the facial nerve of a patient on whom he was performing a routine facelift. The thirty-seven-year-old Clark was distraught not only for the immediate consequences for his patient—essentially a paralyzed half of a forehead muscle, leading to an awkward looking asymmetry—but also for his lack of options in solving the problem. Clark, however, knew of Alan Scott's previous work on using botulinum toxin to eliminate strabismus and blepharospasm, and he reached out to Scott for a word of collegial advice. The senior doctor welcomed Clark's questions and aided Clark in attaining botulinum toxin for injection into the forehead muscles of his patient to alleviate the asymmetry she was experiencing. Clark's unfortunate accident became the first time botulinum toxin was applied to solve a cosmetic issue.[15] One must applaud Clark's honesty and the inherent difficulty in admitting he caused the problem, along with the steps he took to rectify it.

Although Richard Clark was the first to apply botulinum toxin to fix a cosmetic flaw, it would be the duo of Jean and Alastair Carruthers, married ophthalmologists hailing from Vancouver, British Columbia, who pioneered the use of botulinum toxin as a cosmetic enhancer in the manner it is now used today.[16] The couple fostered the development of many of the more practical cosmetic techniques, including the injection of botulinum toxin into the muscles of the face and brow to ameliorate glabellar lines (frown lines), forehead lines, crow's feet (wrinkles that form on the outer corner of the eye), and other wrinkles as well as asymmetric brow height.[17] Soon after the Carrutherses began their work, it was learned that a massage at and around the injection site after the injection of botulinum toxin for cosmetic purposes aids the even distribution of the toxin for a more pleasant aesthetic effect.

Botulinum toxin is measured not in grams or ounces but in units, with a unit of botulinum toxin type A (i.e., Botox) defined as the amount of toxin necessary to kill 50 percent of eighteen- to twenty-gram Swiss Webster mice, a specific breed of mouse commonly used in scientific research. The lethal dose for a human is close to 3,000 units for a seventy-kilogram (roughly 155-pound) person, so the 100-unit vials in which Botox is typically shipped are extremely dilute and should be of no worry to those looking forward to their next injection.[18] A typical injection of botulinum toxin type A for cosmetic use is twenty-five to thirty units. The FDA granted Allergan the right to market Botox as having the ability to reduce the appearance of severe glabellar lines in 2002. In 2013, the FDA would approve the use of Botox for the elimination of crow's feet.[19] By 2013, sales of Botox would top $2 billion and account for more than one-quarter of Allergan's annual revenues.[20] The Botox craze and the era of women (and men) becoming enamored with an age-defying drug stolen from the toxic residue of a ubiquitous bacteria had begun. As a result, the value of Allergan skyrocketed, with the American biopharmaceutical company AbbVie Inc. acquiring Allergan for $61.7 billion in 2019.[21]

A second formulation of botulinum toxin type A, Dysport, would gain FDA approval in 2009 but appear several years earlier in Europe due to its manufacturing base in France. A second serotype of botulinum toxin, type B, would be marketed as Myobloc beginning in 2009 as well.[22] While Dysport often serves a similar cosmetic role to Botox, Myobloc is specified for use in the treatment of cervical dystonia and chronic sialorrhea (excessive drooling).[23]

How Botulinum Toxin Works

The mechanism by which botulinum paralyzes muscles is understood quite well. It does, however, involve some tricks of biochemistry taking place after the botulinum toxin is secreted before anything is able to happen. Botulinum toxin is secreted as a single protein chain that is then cut into two different parts: a large subunit, the "heavy chain," weighing in at around 100,000 daltons, and a small subunit, the "light chain," taking up the remaining 50,000 daltons of the protein's mass. The two chains are linked by a sulfur-to-sulfur bond called a disulfide bond. The heavy chain acts first, binding to the outside of a nerve ending, with the heavy and light chains then taken into the nerve cell. Once inside, the light chain enters the cytoplasm of the cell and begins to take apart a group of proteins called SNARE proteins that "ensnare" packages of acetylcholine in the nerve cell and aid in their release by the nerve. With the SNARE proteins taken apart and left unable to function, the vesicles of acetylcholine cannot leave the nerve cell and share their signal. Acetylcholine is a neurotransmitter used by nerve endings to signal muscles to contract, and without acetylcholine being released, the nerve is functionally shut off, and the muscle becomes flaccid.

In time, new SNARE proteins are made by the cell, and the effects of botulinum toxin wear off, allowing the nerve ending to transmit acetylcholine once more. This is the reason why botulinum toxin injections must be repeated every three months or so in most therapies.

Botulinum Toxin for Migraines

Migraines are defined as recurring, throbbing head pain that lasts for anywhere from a few hours to multiple days, with these headaches accompanied by aura—a variety of visual oddities that usually arise before the migraine itself—and nausea as well as light and sound sensitivity.[24] The recurring nature and length of the migraine makes for a heavy burden on the individual. During an attack, the baseline aspects of day-to-day life often feel insurmountable, let alone work, family, or school responsibilities.

Reports in the mid- to late 1990s surfaced of patients receiving botulinum toxin injections for aesthetic reasons and also noting a decrease in headaches and in muscle pain. Enough reports surfaced to inspire clinical trials to examine whether botulinum toxin type A would be successful in curbing migraines, with the FDA granting approval for the injection of botulinum toxin type A into superficial cranial musculature in 2010.[25] These injections are a prophylactic treatment performed in hopes of eliminating future migraines. There is a significant delay between the moment of injection and a reduction in migraines, with therapeutic effects often not observed for twelve weeks and a meaningful reduction in headaches not seen until more than a year later.[26]

Unlike botulin toxin's action on nerves, the exact mechanism as to how botulinum toxin prevents migraines is unknown, but there are several speculated methods. When injecting botulinum toxin for migraines, twenty-five units of botulinum toxin type A (a little less than the typical cosmetic dose) are injected into the glabellar, frontalis, and temporalis muscles. There are four theories regarding the mechanism by which botulinum toxin prevents future migraines. The first suggests that the toxin acts on the muscle and prevents pericranial muscle spasms, which pull on the bones of the skull and the sutures, the fibrous joints between the bone plates of the skull. When these spasms occur, it is thought that the intercranial pressure in cerebral blood vessels changes, and this may lead to a migraine. Therefore, preventing these spasms by this mechanism would prevent future occurrences of migraines. The second proposed mechanism is similar to the first, suggesting that injections of botulinum toxin prevent the contraction of cranial muscles altogether. The third proposed mechanism suggests that botulinum toxin prevents acetylcholine from entering the blood vessels and that this prevents the dilation (i.e., the widening) of blood vessels near the injection site, which would place additional pressure on the surrounding area and possibly spark a migraine. The last of the proposed mechanisms is the least defined, hypothesizing that just as botulinum toxin blocks the release of acetylcholine, the toxin may block the release of a currently unknown neurotransmitter that triggers the occurrence of a migraine.[27]

Botulinum Toxin as a Band-Aid

In the years since the first cosmetic application of botulinum toxin, scientists and physicians have found a plethora of applications for the molecule, ranging from allergic rhinitis to writer's cramp to decreasing perspiration in the palms, armpits, and feet due to disorders of the autonomic nervous system. Each of these therapeutic uses involves injecting various amounts (starting as low as two units of botulinum toxin in allergic rhinitis, a fraction of the amount necessary for cosmetic use) into the afflicted area, with the mechanisms by which it works for each under investigation akin to the search for a mechanistic explanation for botulinum toxin's ability to fight migraines.[28] Botulinum toxin is also seeing use in the treatment of bladder disorders resulting in incontinence, prostatitis, and constipation due to uncontrolled tightening of the rectal sphincter.[29] Each of these uses shows the versatility of botulinum toxin in manipulating the muscles of the body and has made for plenty of work for interested scientists and clinicians, much of which involves discerning the minimum dose needed to obtain the desired result. The dosage is monitored not only for toxic effects but also to minimize the creation of antibodies to the toxin that render it useless.

Botulinum toxin is also being investigated for use in children with cerebral palsy to alleviate spastic paresis, wherein the muscles are overactive and tighten, preventing the normal flow of fluid through the muscle as well as impairing the limb. The injection of botulinum toxin type A into a muscle exhibiting spastic paresis quickly reduces the muscle tone, with movement and motor function improving when combined with classical physical therapy. Studies into the use of botulinum toxin in patients with cerebral palsy began as early as 1993, with it now deemed effective to inject the toxin into cerebral palsy patients as young as two years old in hopes of preventing surgery and improving limb movements in later childhood.[30] Botulinum toxin has also been applied experimentally to treat depression, lower back spasms, symptoms of Parkinson's disease (including its very commonly observed dystonic clenched fist), and premature ejaculation.

In 2019, the American Society of Plastic Surgeons reported nearly 7.7 million injections of botulinum toxin type A for cosmetic use alone in the United States.[31] Despite the large number of injections made for cosmetic use let alone other applications, using botulinum toxin for any of the indications noted is not cheap, with prices often starting at several hundred dollars for the initial session. The need to reinject with botulinum toxin every three months or so to continue seeing positive results quickly makes the continued application of the drug costly.[32] Part of the cost is inherent in the drug's nature as a protein, which must be expressed by *C. botulinum* and safely purified and transported before it reaches the desired point of use.

Attempts to Weaponize Botulinum Toxin

Although humankind has found a variety of therapeutic means by which to apply botulinum toxin, it is still susceptible to misuse as an agent of harm. In 1990, the Japanese cult Aum Shinrikyo attempted to spray a solution of botulinum toxin from a truck in the immediate vicinity of the Japanese parliament in hopes of cutting the head off of Japan's government by killing as many as possible. The attack was unsuccessful.[33] The name Aum Shinrikyo loosely translates to English as "the supreme truth," with the group consisting of a dedicated number of educated individuals following the precepts of Buddhism mixed with Hinduism and Christianity in the 1980s before things went awry.[34] Aum Shinrikyo tried again to dispense *C. botulinum* in 1993 during the wedding of the crown prince and was once again met with failure. The cult abandoned the weaponization of botulinum toxin and turned to sarin gas for its next ploy, culminating in their March 20, 1995, subway attacks and the deaths of twelve and injury of 3,800.[35]

Despite the failures of Aum Shinrikyo, botulinum toxin is still considered a real and present danger in the bioterrorism community, at least according to a 2001 report of the Working Group on Civil Defense, a panel of the Center for Civilian Biodefense Studies at the Johns Hopkins School of Public Health. Part of the reason for the fear? The relative ease of access—it is found within the soil after all—combined with the simplicity of growth and its deadly nature.[36] Extracting botulinum toxin from *C. botulinum*—the part of the job Aum Shinrikyo likely failed to do properly—is a difficult task but could be accomplished by a dedicated rogue group or nation-state with sufficient financial backing and biochemical know-how. Spreading botulinum toxin through contamination of the food supply with either the toxin itself or *C. botulinum* would be a "dirty" pathway for dissemination, with a more refined and deadlier route coming through the aerosolization of the toxin. Compounding the deadliness of contamination would be the inherent difficulty in successfully diagnosing botulinum toxin exposure, as cases of botulism are quite rare, with the United States reporting fewer than 200 a year.

One foundational aspect plays a role in preventing the use of purified botulinum toxin as a weapon. The first is the fragility of the purified product. For example, a vial of Botox, if shaken too much, can render the toxin useless.[37] The delicate nature of the toxin would make it quite difficult to successfully spread, and it is possibly why Aum Shinrikyo failed when it went with the shotgun approach of spraying members of the Japanese parliament.

Prior to the Gulf War and during Iraq's period of experimentation with chemical and biological weapons, the government of Iraq created large amounts of botulinum toxin but was unsuccessful in weaponizing the threat.[38] Japan's biological warfare unit, the infamous Unit 731, reportedly fed cultures of

C. botulinum to prisoners of war. The United States sought to weaponize botulinum toxin during World War II, partially out of a fear Germany was doing the same thing.[39] It is unknown whether the United States was successful, although a story circulates of a U.S. Office of Strategic Services plot to outfit prostitutes sympathetic to the Allied struggle with pinhead-sized gelatin capsules containing botulinum toxin that could be easily concealed behind the ear or in the scalp. Should one of these ladies come in contact with an Axis officer, it was hoped they might drop a little something extra in the evening's drink.[40] Regardless of the veracity of this tale, the United States feared the use of botulinum toxin in World War II enough to prepare more than 1 million doses of a botulinum toxin vaccine in preparation for the D-Day invasion of Normandy on June 6, 1944, harkening back to the stockpiling of penicillin for the exact same event.[41]

Botulism in Infants and Inmates

Small outbreaks of botulism have happened in the prison population over the past couple of decades. Why? The prevalence of internet recipes for prison wine (affectionately named "pruno") allows prisoners to make alcoholic concoctions using materials from the cafeteria and the correctional facility's commissary. Five inmates at a California correctional facility in 2004 and 2005, eight inmates in Utah in 2011, and an unidentified number in the Arizona prison system in 2013 contracted botulism by imbibing pruno. Pruno is made by mixing potatoes, any readily available fruit like apples or peaches or even cups of fruit cocktail, a number of ingredients readily obtained from the prison commissary, and warm water and allowing the concoction to ferment to create a drink that often "smells like baby-poop."[42]

However, one common ingredient at all three prisons was behind the presence of botulism: potatoes. As *C. botulinum* is readily found in the soil, it would not be out of the realm of possibility for spores of the bacteria to reside on raw potatoes and find their way into a batch of pruno. *Clostridium botulinum* could also be found on baked potatoes, with laboratory studies showing the bacteria to be hearty enough to survive the baking process.[43] The warm but not too-hot temperatures used to ferment pruno are perfect for the growth of *C. botulinum*. Additionally, inmates often use plastic bags to brew the pruno in, providing a classic anaerobic environment for *C. botulinum* to secrete botulinum toxin into the fermenting wine. All in all, the pruno creation process is a nearly perfect environment for the growth of *C. botulinum* and the secretion of its deadly toxin.

The presence of *C. botulinum* on potatoes has led a number of prison facilities to outlaw the use of potatoes in the kitchen to prevent possible contamination of the already illicit pruno.[44] Unfortunate sickness aside, some pretty disgusting mental images are yielded from the forensic examinations that took place to discover the cause of botulism in the inmates, including a still-moist sock that Utah investigators announced tested positive for botulinum toxin. Why? It was used to filter the pruno.[45]

A second circumstance rife with cases of botulism is the infection of an open wound in a low-oxygen environment. Wound botulism can occur in users of injectable illicit drugs and, in particular, among users of black tar heroin. The gummy substance black tar heroin is sometimes contaminated with soil containing the bacteria *C. botulinum*. *Clostridium botulinum* often survives the heating portion of the preparation process for the drug. The practice of "skin popping," the injection of black tar heroin just below the surface of the skin, also provides a low-oxygen environment for *C. botulinum* to flourish or make an already putrid wound far worse.[46]

Infants are also at a heightened risk for botulism due to the relative "newness" of their gastrointestinal tracts and the fact that their bodies have not had enough time to build up natural defenses to *C. botulinum*. For this reason, it is advised against infants eating honey, which can contain spores of *C. botulinum*. The bacteria can live long enough in their gastrointestinal tracts to secrete botulinum toxin, possibly leading to botulism in the infant. Older children and adults do not have this problem, as their gut floras are developed enough to eliminate small amounts of *C. botulinum* before any toxin can be expressed.[47]

Fighting Botulism

Although modern cases of botulism are few and far between, a botulism antitoxin exists for use in patients with severe cases of botulism. It does not stop already occurring paralysis but does prevent the spread of the toxin, with paralysis reversing in time as the proteins cleaved by the light chain of botulinum toxin are created again by the cell. The antitoxin Botulism Antitoxin Heptavalent (A, B, C, D, E, F, G) - (Equine), abbreviated BAT, is derived from horses and is a mixture of antibodies targeting all seven serotypes of botulism toxin. As such, it is successful in protecting adults displaying signs of botulism from further damage incurred due to the seven known types of botulinum toxin.

Infant botulism, due to the low number of cases annually, is considered an orphan disease, with no treatment until the antitoxin BabyBIG was approved by the FDA in 2003. The antitoxin is derived from humans and was created by an unusual source—the California Department of Public Health. Despite its altruistic source and orphan drug status, BabyBIG is not cheap, coming in at more than $45,000 per vial.[48] Studies have shown that in outstanding cases, equine botulism antitoxin could be applied to infants, particularly in situations where the very expensive BabyBIG is not available.[49]

From blood sausage to pruno, humankind, much to its detriment, has found a variety of ways to support the growth of *C. botulinum* and its production of botulinum toxin. Thankfully, a few keen minds looked at the toxin as an opportunity to heal, revolutionizing the cosmetic industry in the process while providing relief to those suffering from debilitating muscle contractions and migraines. This would make for a bizarre journey for any medication let alone one that breaks the mold of traditional drug characteristics like botulinum toxin.

CHAPTER 12

Coal Tar

Pentacene, chrysene, and coronene, which you can see here, are examples of polyaromatic hydrocarbons. A sample of coal tar contains a plethora of different polyaromatic hydrocarbon molecules. *Structure generated by author in ChemDoodle v 11.5.0, using information supplied by the National Institutes of Health*

The name "coal tar" sounds a little gross and immediately brings to mind images of asphalt being used to pave roads and its awful smell. Coal tar, although not the best-smelling item in the world, instead provides an emotionally uplifting confidence boost to many suffering from a variety of skin diseases. The substance is neither ingested nor injected but rather is used as a topical agent for skin care and to fight psoriasis and dandruff. Coal tar is also very different from the other medicines in this book, as it is not a single, known molecule but rather a mixture of thousands of molecules, many of which are still a mystery. What is it about coal tar that leads to its varied dermatological uses, and just how do we even obtain the substance?

Just What Is Coal Tar?

Coal tar is a mixture of more than 10,000 different organic compounds, many of which have not been identified. These various organic molecules are made when coal, which is almost entirely carbon with trace amounts of hydrogen, oxygen, nitrogen, and sometimes sulfur, is heated to extreme temperatures in the absence of oxygen. The ensuing vapor is then allowed to condense and yield this bounty of molecules formed through millions of different individual reactions.[1] Among these 10,000 molecules are polyaromatic hydrocarbons—compounds constructed of fused carbon rings that allow their electrons to flow freely from one ring to another—a beleaguered cadre of chemicals due to many of them being carcinogenic. Their cancer-causing nature rears its head when they are acted on in the body by enzymes that slightly change the polyaromatic hydrocarbon into a form that can interact and bind with DNA, leading to mutations and altering a litany of cellular processes. This, to make a long story short, is not good.

Two main versions of coal tar are available for medical use: a stronger pix lithantracis, which is pungent smelling and usually restricted to hospitals, and an extraction of pix lithantracis in alcohol that yields liquor carbonis detergens for over-the-counter and home use (although it too is still quite strong smelling).[2] It is highly bizarre that this by-product of coal distillation on the way to making coke would have any useful effect on the human body. Distillation is the act of purifying a material, often a liquid, through heating and cooling, and if you took a course in organic chemistry, you no doubt distilled many a mixture in your labs. Coal is distilled to create coke, a more purified form of coal that is almost entirely carbon. Heat is applied to soften the coal, with the coal eventually changing to a liquid form as the temperature rises before the product cools and you are left with the solid, high-carbon portion known as coke and the remaining liquid, coal tar. Coke can then go on to be used in a number of industrial processes, including the production of molten iron and steel.[3]

Deodorizing Feces and Fighting Scabies

Historically, one of the contributing factors that led to the plethora of uses for coal tar was the fact that there was just so much of it. Lighting houses and street lamps in England with coal gas was popular as early as 1794, and as its popularity grew and gas lighting reached the shores of the United States, so did the amount of coal tar on hand. One of the first uses for all of this excess coal tar was as a disinfectant for manure to be used in making fertilizer. Going one step farther in the chemical process of attaining coal tar, the distillation of coal tar allows one to collect phenol, a combustible clear liquid used as a disinfectant in the mid-nineteenth century. Phenol saw use as a wound dressing, but its most popular use would be quite dirtier in nature. Phenol from coal tar was mixed with calcium salts to create the commercial product McDougall's Powder, which had the unglamorous primary task of deodorizing sewage.[4] This was all well before widespread sewer systems in cities, and anything successful in removing the smell from the piles of animal and human waste prevalent would be quite the moneymaker.

Another one of the first commercial products to include coal tar came out of England in 1860—Wright's Coal Tar Soap (it was also referred to as the much more scientific-sounding Sapo Carbonis Detergens, but I think you can guess which name won out in popular opinion). The orange soap was made from liquor carbonis detergens and claimed antiseptic properties.[5] Wright's Coal Tar Soap was created by William Valentine Wright, a druggist (an older term for a pharmacist) who gained no small amount of fame and success by producing a brand of nonalcoholic communion wine and in doing so further driving the dividing line between drinking alcohol and the church.[6] Wright's Coal Tar Soap would be a runaway hit, with the soap still sold in Europe today albeit in a modified form. Unfortunately for Wright, however, his antiseptic soap played no role in stopping the infection that took his life, as the inventor came down with a case of erysipelas at the age of fifty-one. Erysipelas is a scarcely fatal superficial skin disease stemming from an infection with streptococci or *Staphylococcus aureus* that in rare cases turns septic with the patient developing the "yes, it is as bad as it sounds" necrotizing fasciitis, which rapidly eats through the body's soft tissue.[7] The irony of the inventor of an antiseptic soap dying of a superficial skin disease is not lost.

Due to current European Union constraints, coal tar is no longer an ingredient in the modern form of Wright's Coal Tar Soap, although the manufacturers take steps to maintain the distinct coal tar odor. The new formulation, sold as Wright's Traditional Soap, maintains an element of antiseptic effect by substituting tea tree oil for coal tar as the active ingredient.[8]

Formal reports of the use of coal tar oil for medicinal purposes date back to the middle of the nineteenth century, with an 1861 letter to the *British Medical Journal* lauding its use to fight scabies. Scabies is an infestation of *Sarcoptes scabiei* on the skin with the mites burrowing downward and leading to severe itching and discomfort, with coal tar coming into use when the traditional sulfur-based treatment was unavailable.[9]

A second early, crude use came from turn-of-the-twentieth-century millworkers covering cuts or even dipping the stubs of missing fingers into a crude dye made of coal tar, swearing that the coal tar mixture aided the wound-healing process and provided an antiseptic barrier.[10] Coal tar was even used as a treatment for one of humankind's great vacation ruiners—bedbugs. Contact with coal tar itself or high concentrations of the vapor are lethal to bedbugs, leading those infested with *Cimex lectularius* to spray the walls, ceilings, floors, and bed itself with coal tar oil, only to quickly learn it would not only stain the surrounding area but also destroy paint and linoleum. This led to an earnest effort to determine a less messy way to kill bedbug-infested furniture. What researchers came up with is a "van fumigation" method wherein the infested furniture is loaded into a van, with a combination of coal tar vapor and high temperatures used to kill the bedbugs in a matter of hours while preventing the destruction of your bedroom.[11]

Further distillation of coal tar yields the chemical quinoline, which was used for pain relief and to quell fevers. The distillation of coal tar made for a simplified industrial process by which to create large amounts of quinoline, so much so that quinoline, even with its disgusting taste and smell, was used to fight typhoid fever and rheumatism until the dawn of the twentieth century. Part of what led to the discontinuation of quinoline was the rise of aspirin during the late 1800s and early 1900s, which we concerned ourselves with in an earlier chapter. Aspirin performed in a similar or more successful manner than quinoline without the poor taste or another particularly negative side effect—turning one's skin blue. Ingesting quinoline in a small segment of the population led to cyanosis due to an impediment of the oxygen supply in the body.[12]

In time, coal tar would come to be used for what it is now associated with fighting dandruff and the effects of psoriasis. Dandruff is never wanted, yet it makes an appearance in nearly all of our lives, rearing its head when we are adorned in black sweaters to earn the name from which it is taken. The word dandruff is a combination of the Anglo-Saxon words "tan" and "drof," yielding the phrase "dirty tetter." Here the word "tetter" refers to an underlying skin disease that may or may not be at play.[13] Dandruff can be the result of a drastic drop in humidity and annual battles with dry skin, but it can also be a manifestation of seborrheic dermatitis, a long-term skin disease that manifests as itchy, scaly, and greasy skin. A plethora of "medicated" coal tar–containing shampoos are on the market, with many available over the counter. Coal tar works to suppress dandruff by allowing the scalp to shed the dead skin cells residing on the top layer of skin and also inhibit their replacement by slowing down the proliferation of skin cells in the applied region.[14]

Psoriasis is an adjacent skin condition but one with more far-reaching effects. In psoriasis, skin patches, often red and scaly, form along the scalp and elbows. These are frustrating to the patient on a personal level, as they radically alter one's appearance, not even taking into account the patches' ability to crack and bleed at times. The exact chemical mechanism as to how coal tar provides relief is unknown. This makes a lot of sense, especially when you consider that a large portion of the problem in solving the mechanism comes in teasing out the culprit in the legions of individual molecules that make up coal tar. Studies have shown that coal tar activates the aryl hydrocarbon receptor and restores the production of filaggrin, a skin barrier protein, which may hold part of the key as to the mechanism.[15] Due to the sheer number of molecules involved in coal tar, however, we may never have an exact mechanism for how coal tar works as we do with many of the other drugs in this book such as iproniazid or acetylsalicylic acid.

It is believed that coal tar aids in the treatment of psoriasis, like dandruff, in slowing down the growth of skin cells in the general region. It is also thought that coal tar aids in altering the differentiation of cells in the outer layer of the skin and how the cells express keratin (a protein associated with nails and hair) in patients exhibiting psoriasis.[16] Coal tar may also express an anti-inflammatory effect and decrease itching in those with psoriasis, but this is harder to quantify. Coal tar appears to have a similar effect on those with atopic dermatitis, a condition better known as eczema.

Early Concerns about Coal Tar and Cancer

As mentioned, coal tar is created during the destructive heating of coal and capturing the vapors that are emitted during the process and then cooling them. What you are left with is a dark black fluid composed of more than 10,000 different molecules.[17] Occupationally, this is a very dangerous job and carries a high cancer risk, with the link between the breaking down of coal to obtain coal tar and cancer known since a retrospective report in England in the 1930s. The study found more than 700 individuals within its purview to exhibit skin cancer. A longer-term German study performed later in the century found similar results in an increased incidence of skin cancer. Along with skin cancer, an increase in scrotal, buccal cavity (the parts of the body in and around the mouth), and pharyngeal cancers were observed in those working in this environment.[18]

Aware of the occupational risks of cancer in the presence of coal tar, scientists took up the mantle of determining whether they could reproduce the event in a laboratory setting. They turned to their favorite test subjects, mice, painting sections of the skin of twelve mice twice weekly with a solution of coal tar in 20 percent ethanol. In the forty-one-week study, seven of the twelve mice developed papillomas (benign skin tumors). Tumors on four of the mice eventually became malignant as time passed in the trial. Laboratory evidence like this carefully crafted experiment was a blow to the use of coal tar, with the authors of the study hypothesizing that the length of time of use played a role as to whether cancer would eventually develop and ending with a plea to the reader to avoid the long-term use of the substance.[19]

The fact that some of the compounds in coal tar play a role in the manifestation of cancer is not out of the realm of possibility, as coal tar is a mixture of no fewer than 10,000 compounds, and of those identified, many are polyaromatic hydrocarbons. The link between the long-term use of coal tar or taking part in the process of manufacturing it for an extended period of time and cancer is disturbing and has led some dermatologists to all but abandon its use.

Pushback for Coal Tar–Containing Products

In 2001 and intermittently in the years following, the National Psoriasis Foundation has fought for the availability of over-the-counter products containing coal tar, like shampoo, soap, and ointments. Such products are highly unlikely to increase the risk of cancer, especially when used in the short term and in small amounts. In 2001, the State of California threatened lawsuits against manufacturers of coal tar shampoos and products to have warning labels put on the products. Why? Under California's Proposition 65, which was enabled in 1986, items containing chemicals known by the state to be carcinogenic or to contain reproductive toxins require warning labels.[20] The State of California eventually won, and over-the-counter coal tar–containing products must now carry a warning. Despite the concern of the state, the U.S. Food and Drug Administration (FDA) has reiterated that dermatological use of over-the-counter products containing coal tar does not prompt an increased risk of skin cancer and is viewed as safe and effective.[21] An exception here is coal tar–based hair dyes, which, in the opinion of the FDA, could cause an increase in the incidence of skin cancer per data from animal studies.[22]

Should people be afraid to use coal tar? Probably not. A 2010 retrospective study of 13,200 individuals over multiple decades showed the use of coal tar treatments in patients with psoriasis or eczema to not correlate with an increase in cancer risk when compared to the normal risk of cancer that occurs in one's lifetime.[23] As such, coal tar should be within the toolbox of anyone suffering from these dermatological disorders.

CHAPTER 13

Minoxidil

The structure of minoxidil.
Structure generated by author in ChemDoodle v 11.5.0, using information supplied by the National Institutes of Health

Balding. It's a natural part of aging for many men and women and a major impairment for many, as it drastically changes their physical appearance. During the twentieth century, bald men tried a variety of means to spur hair growth, ranging from the comical—standing on one's head to improve circulation to the cranium—to the life changing—taking hormone cocktails that overwhelmingly failed to produce hair growth yet left them with a depleted sex drive and enlarged breasts.[1] If they only knew that real hair growth at minimal cost would be available in the last decades of the century in the form of minoxidil.

An Unusual Side Effect
Leads to a New Medication

Minoxidil, or 2,6-diamino-4-piperidinopyrimidine-1-oxide as it is named in chemical longhand, started out as an analogue derived from N,N-diallylmelamine. Thankfully, N,N-diallylmelamine has a much more palatable, shortened name: DAM. Scientists at the Upjohn Company were testing the impact of DAM on stomach ulcers, which DAM was eventually shown to be ineffective against. Researchers, however, did take notice as DAM resulted in a precipitous drop in blood pressure in patients.[2] The next step taken was to modify DAM, which scientists did by the creation of DAMN-O, which Upjohn ushered into clinical trials for its hypertensive properties in 1961. DAMN-O worked in relieving blood pressure, but as we will see with oral minoxidil, its side effects were pronounced, with edema and heart failure occurring in some patients. DAMN-O also performed poorly in animal studies, as it created hemorrhagic lesions in the right atrium of the hearts of dogs.[3] Because of these dire side effects, DAMN-O was shelved, but a number of analogues synthesized, with 2,6-diamino-4-piperidinopyrimidine-1-oxide, the fabled minoxidil, being one such analogue. At 209 daltons, minoxidil looked to be a reasonable drug candidate. It was extremely successful in treating hypertension, and while it showed similar side effects to DAMN-O, researchers and physicians pressed on for further testing. The U.S. Food and Drug Administration (FDA) granted emergency use approval for oral minoxidil in 1971. The approval aimed at treating patients with severe hypertension but only for a treatment period of two weeks. The rationale was that this was long enough for the positive effects on hypertension to be observed but not so long to allow the dramatic negative side effects, particularly fluid retention, to set in.[4]

Studies of the effect of minoxidil on hypertension were under way under the oversight of Charles A. Chidsey of the University of Colorado in 1971, when one first-year medical resident, Paul Grant, noticed the particularly odd plight of a woman taking the drug. The patient, in her mid-forties, had already experienced two strokes and was the perfect candidate for the experimental use of oral minoxidil at that time, as anything that would relieve her body of the exorbitant threat of hypertension would be well worth any side effects. She and her physicians, however, did not plan for one particular side effect: the growth of hair all over her face and the increased growth of hair on her head and legs. This was all coming from a woman who admittedly rarely if ever had to shave her legs.[5] Paul Grant mentioned this unusual side effect to Guinter Kahn, the head of dermatology at the University of Colorado School of Medicine, who immediately jumped to the million-dollar (well, billion-dollar in this case) application—what if we could apply this to people's heads? Guinter Kahn, who would champion the future drug, was thirty-seven at the time, a German Jewish emigrant who fled the Nazis with his family and began a new life in Omaha, Nebraska, at the age of four in 1938.[6]

The first step to using minoxidil to restore hair would be turning minoxidil into a topical medication. Why? So that its actions could be concentrated on a desired portion of the body instead of the runaway hair growth seen in Paul Grant's patient who was taking the oral form. Kahn surreptitiously acquired some minoxidil powder from Chidsey's study and began making 1 percent solutions of dissolved minoxidil in ethanol and propylene glycol. The latter is a common vehicle for topical medications and a wise, informed choice on the part of Grant and Kahn. Kahn set out to test the solution, much like the drug discoverers of old, on Grant and himself along with two other people, one a secretary in his department and the other an unnamed medical resident.[7]

Grant and Kahn were working with a fraction of the concentration of the current formulation of minoxidil, and after results never showed for Kahn, enthusiasm waned. This was the Vietnam era, with Paul Grant serving as a member of the U.S. Army Reserves and obligated to two-week stints of active duty at nearby Fort Carson. One morning in the barracks, Grant pulled back the one-inch bandage covering a spot of skin on his right arm and finally saw the result Kahn and himself were vying for. Waiting underneath the bandage was new hair growth—and not only new hair but healthy, dark hairs. The secretary and unnamed resident saw similar results, with Kahn's daughter commenting after his death that Kahn appeared to be allergic to minoxidil and thus never got to make use of the discovery to treat his own receding hairline.[8]

On returning to the University of Colorado, Grant quickly told Kahn about the surprise on his right arm. Minoxidil not only grew vellus hair, the light "peach fuzz" that covers the legs, stomach, and arms of some people, but also promoted the growth of new terminal hair. Terminal hair is the dark, coarse, thick hair associated with facial hair as well as the hair of the scalp, underarms, and pubic area.[9] The duo informed minoxidil's creator, the Upjohn Company, about their serendipitous discovery informed by the side effects of a female hypertension patient, with Upjohn immediately filing for a patent for the use of minoxidil as a hair growth aid in 1971. Upjohn also filed a second set of papers, but these held devious intent. The company snitched to the FDA about Grant and Kahn making use of humans in pharmaceutical experimentation without proper authorization.[10]

At this point, it appears that the Upjohn Company sat on its knowledge of minoxidil's hair restoration properties for several years. Why? It is not exactly known, but it is possible that the company wanted to maintain a serious focus and appearance by steering clear of the hair growth industry.[11] Such a foray was regarded about as well as peddling snake oil in the nineteenth century. It was well known within the medical community that minoxidil's extended oral use led to the excessive growth of hair all over the body, a condition known as hypertrichosis. The 1980 case report "Reversal of Baldness in Patient Receiving Minoxidil for Hypertension" in the *New England Journal of Medicine* blew the lid off of minoxidil's much-desired side effect within the medical community. When word of Upjohn's secret made rounds through its home city of Kalamazoo, Michigan, the company was at no loss for balding volunteers looking to take part in a clinical trial. The Upjohn Company's primary concern at this time, however, was getting minoxidil to market not as a hair restoration product but as an antihypertensive. Once it succeeded in this arena by bringing oral minoxidil to market as Loniten in 1979, Upjohn changed course and started the exploration of topical minoxidil for hair growth. In the meantime, off-label prescribing of oral Loniten was a popular method used by physicians to spur hair growth in patients despite the additional side effects of oral minoxidil.

During the intervening years, a number of clinical and scholarly trials were being carried out to test just how well minoxidil prompted hair growth. Early hair growth tests—the nonhuman kind at least—took place using stump-tailed macaques, which naturally exhibit thinning hair and balding in some cases. These were successful, opening up a route to use minoxidil on humans or, from a different point of view, to retrace the steps Kahn and Grant had already trod but this time with full authorization in hand. An early 1983 study looking at the effect of topical minoxidil on alopecia areata (a circular, patchy loss of hair) using crushed tablets dissolved to make an ointment and lotion in the manner of Kahn and Grant showed definitive results in sixteen out of thirty patients and marginal results in another six.[12] Minoxidil was also taking on the order of six weeks to start working, not a very long time, especially considering the lengths people were going to prior to the availability of a medicinal option. In sum, the outlook was quite hopeful.

Clinical trials of a topical minoxidil treatment started with a very small amount of minoxidil—a 0.01 percent solution—and ramped up to 2 percent.[13] Substantial hair growth was seen only in the patients taking 1 to 2 percent solutions, further displaying the role luck played in Kahn's creation of a 1 percent solution to test on Grant and his cadre. Any lower, and they would have missed out on the hair growth effects of the drug.

In 1988, the FDA finally approved Upjohn's minoxidil for prescription hair growth under the name Rogaine. Upjohn's initial name of choice was Regain, but the FDA felt the company was overstepping its bounds, as the drug did not work in every patient.[14] In February 1996, an over-the-counter version of minoxidil as well as a variety of generics were approved, bringing the price of topical treatment down to as low as $30 a month.[15]

To this day, androgenetic alopecia (male-pattern baldness) and general fe-male hair loss are the only two FDA-approved uses, with all other applications of topical minoxidil to hair loss considered off label.[16] "Alopecia" is the medical term for hair loss, with several types of alopecia designated by physicians. Andro-genetic alopecia is the most common form of hair loss and is also known as male-pattern baldness. Minoxidil is often used to treat three other types of alopecia: alopecia universalis, which is the complete loss of all body hair; alopecia areata, which is the loss of round, patchy parts of the scalp; and alopecia totalis, which is the loss of all hair on the scalp. Topical minoxidil is available as a lotion and a foam. Oral minoxidil is still not approved for treatment of hair loss.[17]

Upjohn brought a 2 percent minoxidil topical lotion to market initially, fol-lowed by an improvement on its own product with a 5 percent minoxidil foam in 1993. The foam provided less irritation to the scalp and applied areas, with the FDA consenting to the increase in active ingredient.[18] The foam is believed to be the superior delivery method.

In treating androgenetic alopecia (male-pattern baldness), topical minoxidil definitively spurs results compared to a placebo. A 2 percent topical minoxidil solution showed an increase of just over eight hairs per square centimeter, while 5 percent minoxidil nearly doubled the result, coming in at just below fifteen hairs per square centimeter of applied skin. When treating hair thinning in women, studies tend to use 2 percent topical minoxidil, which still results in more than twelve hairs grown per square centimeter.[19] This may not sound like a lot, but compared to the alternative of zero hairs grown per square centimeter, these amounts can be a boon to the confidence of those afflicted with alopecia.

The Wrath of Kahn

It is likely Upjohn's avoidance and attempt to censure the duo of Kahn and Grant left a sour taste in the mouth of Kahn, as Kahn would leave the University of Colorado and start a private dermatological practice in Miami Beach, Florida, in 1974.[20] Moving to Miami Beach from most places in the world is not a lateral move, and Kahn stayed there until he died in 2014, so I am confident Kahn was happy with the change. Whatever happiness the change in locale brought about was mired in courtroom drama, with a fifteen-year court battle ensuing. The battle pitted Kahn and Grant against the Upjohn Company, finally ending in 1986 with Kahn being added to Upjohn's patent for minoxidil, while both Kahn and Grant received royalties from the global sale of Rogaine. These royalties were significant, with the rate estimated to be between 2 and 5 percent on nearly $200 million a year in sales. Despite the windfall, it was just as important for Kahn to be singled out for his work. In 1989, Kahn won the Distinguished Inventor of the Year Award from the Intellectual Property Owners Foundation, a small bit of recognition but by all accounts a recognition for which Kahn deeply yearned.[21]

How Does Minoxidil Work?

Minoxidil interacts with a sulfotransferase enzyme, which is present in hair follicles, with the enzyme modifying minoxidil to its active form, minoxidil sulfate. From here, the mechanism of how minoxidil induces hair grown is unknown. This does not stop scientists and physicians from speculating on it, with their speculations centering on minoxidil's action as an arteriolar vasodilator due to the opening of potassium channels in the smooth muscles where it is applied.[22] Interestingly, taking aspirin interacts with the enzyme sulfotransferase present and inhibits its transformation of minoxidil to the active minoxidil sulfate, diminishing the effectiveness of the drug.[23] Take note minoxidil users.

Minoxidil is not the only drug that causes hair growth, let alone overall hair growth, a condition called hypertrichosis. Diazoxide, a medication used to manage low blood sugar, can also result in hair growth. Both diazoxide and minoxidil are vasodilators, and many feel this tandem result lends credence to vasodilation playing a role in minoxidil's action.[24] Vasodilation could lead to an effect that alters how long a single hair follicle stays in a growth stage or how it moves to and from growth stages. A single strand of hair cycles through three different stages before it is shed. The first stage is a long, multiyear anagen phase, when it is actively growing, followed by a short catagen phase, when the hair no longer grows but converts to its final form. The catagen phase is followed by the telogen phase, when the hair "rests" while a new hair begins to grow underneath it and eventually pushes it out. Rat studies suggest that topical minoxidil shortens the telogen phase and possibly allows for anagen phase–like growth during the telogen stage. Topical minoxidil also lengthens the anagen phase in rats by shifting more quickly to the beginning of the cycle, increasing the overall growth of the hair.[25]

The Different Ways Men and Women Lose Hair and a Gender-Based Price Controversy

Along with the different types of alopecia, men and women lose hair in different ways. Men typically see a patterned hair loss, beginning with a recession of their hairline above the temple, with the telltale "M" shape forming and gradually progressing to baldness as recession continues. Women typically do not experience the receding hairline effect but instead exhibit an overall thinning of the hair, with severe cases leading to thinning to the point where the scalp is readily seen. This thinning in females often coincides with the onset of menopause.[26]

To go along with their different balding patterns, men and women often pay different prices at the pharmacy for minoxidil-containing products. A 2016 study showed that women pay nearly 40 percent more for the same-size container of over-the-counter minoxidil foam, as the cost for the version of minoxidil marketed to women is substantially higher than that marketed to men. Both contain the same amount of minoxidil at 5 percent. Playing a role in this is Upjohn's continued patent protection for the use of the foam in women, thus disallowing the place of a generic on the market.[27] This disparity extends to the topical lotion, which is off patent, where the 2 percent minoxidil lotion marketed to women is priced at the same amount as the 5 percent version targeted at men despite a substantial difference in the amount of active ingredient.[28]

Minoxidil works to express more hair in the eyebrows and beard region as well as the chest, allowing for further aesthetic options. A 2014 clinical trial sponsored by Thailand's Mae Fah Luang University Medical Center was undertaken to fully understand how well a 3 percent minoxidil solution would spur the growth of chest hair.[29] I take it that no Olympic swimmers took part in the study due to an increased amount of body hair's negative impact on fluid dynamics, but for those looking to sport the 24/7 sweater look, minoxidil might be the route to go. Once you start taking minoxidil, a dirty little secret is that you must continue to take the medication, or else all of your regained hair will often dissipate within three to four months.[30]

Topical minoxidil has also been shown to be effective in chemotherapy-induced alopecia. This provides some sense of normalcy for chemotherapy patients, as results showed those using minoxidil to grow hair fifty days earlier than those who did not use minoxidil in one study of patients enduring chemotherapy for four months. Similar results, however, were not seen with those using topical minoxidil during chemotherapy for breast cancer, leading some to suggest an oral minoxidil regimen for hair growth be applied during chemotherapy.[31]

Oral minoxidil is meant to be taken for a short amount of time for those with severe hypertension. Continued dosage of oral minoxidil has a minefield of possible side effects, starting with sodium and fluid retention, which often leads to weight gain, with congestive heart failure a concern in many extreme cases. Oral minoxidil can also cause fluid to build up and be retained around the heart, which can lead to irregularities in pumping. Ischemic heart disease is also a possibility, with the heart needing more oxygen due to the changes in heart rate and cardiac output that come with lowering a patient's blood pressure. In women, polymenorrhea (a shortening of the menstrual cycle, often to twenty-one days) can occur when taking oral minoxidil.[32] Aside from hair growth, the effects of oral and topical minoxidil do not overlap. With topical minoxidil, you do not see the change in blood pressure that made the oral version such a successful antihypertensive, but you also do not see the unwanted and sometimes dire side effects associated with taking minoxidil orally.

Minoxidil in any form can be dangerous to our feline friends. A 2004 report shows topical minoxidil to be extremely toxic to cats, making the hair growth solution a no-go for most feline lovers. The evidence is small: two cats that came in skin contact with topical minoxidil, with the furry friends developing considerably lethargic behavior and having trouble breathing within a day and a half. On further examination, both cats exhibited fluid in the lungs and between the layers of the membrane outside of the lungs. The pair unfortunately passed shortly after coming into contact with the minoxidil sample.[33]

CHAPTER 14

Finasteride

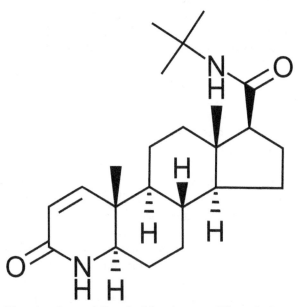

The structure of finasteride. A very different structure from minoxidil, yet both produce the same end result of hair growth. *Structure generated by author in ChemDoodle v 11.5.0, using information supplied by the Royal Society of Chemistry*

Minoxidil is not the only medication that promotes hair growth, as finasteride (also known by the brand name Propecia) has been its main competitor over the past three decades. Finasteride, unlike minoxidil, is available as a pill for those looking to regrow hair, but it is not without its own unique set of side effects. Finasteride began life in a very different manner than minoxidil, however, with the molecule owing its origins to the observations of an astute researcher as she studied a group of hermaphroditic children living on an island.

Güevedoces and a Prostate Enlargement Cure Lead to a Hair Loss Discovery

The journey to finasteride becoming a treatment for hair loss and hair regrowth begins in an unusual place: the Caribbean. There, Cornell Medical College's Dr. Julianne Imperato-McGinley had been studying a group of hermaphroditic children raised as girls. As puberty took its course, the children developed external genitalia, deepened voices, and chest and arm musculature that would be assumed to accompany a growing teenage boy.[1] In the Caribbean, these children were known as "güevedoces," a term that loosely translates to "penis at twelve."[2] Imperato-McGinley gave a pivotal presentation of her data in 1974 at a conference on birth defects, not exactly a crowd looking to tackle the plight of aging men or male-pattern baldness. In Imperato-McGinley's discussion, she noted that children exhibited a mutation in their genetic code that caused less of the enzyme 5-alpha reductase to be present in their bodies, the enzyme that transforms testosterone into the potent molecule dihydrotestosterone, also known as DHT. This lack of 5-alpha reductase suppressed the development of male sexual characteristics until the onset of puberty. In the talk, Imperato-McGinley made the offhand comment that the prostates of the güevedoces remained small. This comment would resonate in a summary of Imperato-McGinley's presentation that circulated around the higher floors of pharmaceutical giant Merck the following year. Merck basic research chief P. Roy Vagelos seized on the connection between small prostates and the lack of 5-alpha reductase activity made by Imperato-McGinley. Vagelos and Glen Arth, a Merck senior scientist, immediately set out find small molecules that would inhibit 5-alpha reductase in hopes of shrinking the prostates of men suffering from benign prostatic hyperplasia, an enlargement of the prostate. Benign prostatic hyperplasia had a massive patient base approaching 15 million individuals at the time, and any molecular hit here would be a potential multi-hundred-million-dollar-a-year drug for Merck.[3] Benign prostatic hyperplasia often occurs as men age and leads to embarrassing difficulties in urination and, in rare cases, infection as the gland surrounds the urethra as it leaves the bladder.[4]

Vagelos and Arth found a proper target for 5-alpha reductase in the 373-dalton small molecule finasteride, which would hit the market as Proscar in 1992. The decrease in prostate gland size was dramatic, averaging about 25 percent.[5] If one stops taking 5 milligrams of finasteride for benign prostatic hyperplasia, however, the prostate enlarges to its previous volume within about three months.[6] The 5-milligram oral form was a success, as prior to finasteride's appearance on the scene, the primary method of shrinking an enlarged prostate was a prostatectomy. (As a man who hopes to one day be rather old, I'll take a pill any day to prevent someone from cutting into my prostate.)

Along the way, researchers discovered a second use for finasteride: hair growth. It turns out that excess dihydrotestosterone in the presence of hair follicles sets the stage for DHT to bind to androgen receptors and the subsequent miniaturization of the follicle. This leads to the prevention of the growth of new hair and its maturation, and thus you get hair thinning and loss. The same molecule used to inhibit the work of 5-alpha reductase to shrink prostates could also be useful to stop testosterone from becoming the more potent dihydrotestosterone and wreaking havoc on the scalp.[7] Just 1 milligram of finasteride a day promoted hair growth and prevented further hair loss according to phase 3 clinical trials of nearly 2,000 men, with this growth maintained by the group of test patients for two years.[8]

Merck, on the heels of Proscar, had a second financial boon on their hands, with minimal additional work necessary. Finasteride, this time packaged under the brand name Propecia, was essentially a 1-milligram tablet form of Proscar. Propecia gained U.S. Food and Drug Administration approval in 1997 for the treatment of androgenetic alopecia, also known as male-pattern baldness.

There is no doubt concerning Propecia and finasteride's popularity. The drug finasteride was prescribed 2.4 million times in the United States in 2020 for hair loss, not to mention the number of times it was dosed for benign prostatic hyperplasia.[9] Former President Donald Trump's last two annual physical exams also show he has taken the drug in the past.[10] There is an argument to be made that finasteride, an oral medication, would have a higher degree of success than a topical medication like its main competitor, minoxidil. Minoxidil must be applied correctly and repeatedly to see growth, but with finasteride, you take a pill on a regular basis and wait. This is Propecia's main selling point: it's a pill, you take it, you grow hair, and you stop losing the hair you've grown. It is also successful in nearly 80 percent of patients who use it for one year.[11] Finasteride's fight against hair loss would not be so easy, however. Once one stopped taking finasteride, hair growth would reverse within twelve months. Finasteride also came with a bevy of side effects, most of them sexual in nature, including gynecomastia (a swelling of male breast tissue), decreased ejaculate volume, and a reduction in penis and testicle size.[12] In some cases, loss of libido and erectile dysfunction were also noted.[13]

Topical Finasteride

Minoxidil, finasteride's primary competitor, is available as a topical solution or a foam. But could we see a topical version of finasteride in the future? Probably not. Finasteride comes with an unusual warning in the form of a pregnancy category X warning, meaning there is a known risk to human offspring from the drug.[14] If a tablet is broken or crushed, a female, regardless of her current pregnancy status, is advised not to touch it, as the finasteride within could be absorbed and lead to abnormalities in a male fetus.[15] As such, women are often advised not to take finasteride. As of the time of this writing, only the oral form of finasteride is approved, be it for benign prostatic hyperplasia or hair growth. This is due to the severe danger a broken tablet of finasteride poses to an unborn child, rendering the likelihood of a topical version (and thus more accessible form) of finasteride unlikely. However, studies have still been performed to determine whether a topical form of finasteride would work to grow hair.

It is probably good to note at this point that minoxidil and finasteride are not the only ways to medically augment hair, as surgical hair transplantation is quite successful in many cases, but the cost is prohibitive or unreasonable for most (especially when compared to $30 a month for over-the-counter minoxidil or the availability of low-cost generic prescription finasteride). Hair transplants persist as an option for those with the financial wherewithal and when both finasteride and minoxidil have failed.

Topical finasteride holds hope, as it would be a way to block dihydrotestosterone at the hair follicles but without the impact on serum testosterone that is seen when taking the medication orally and thus, hopefully, fewer side effects. Initial topical tests took place with a very small volume (one milliliter) and a very low concentration of finasteride (0.005 percent). Successful hair growth was seen in men and women after six months of application with no significant change in plasma levels of testosterone or dihydrotestosterone. This gave hope that the side effects seen with oral finasteride could be avoided due to the lack of change in plasma hormone level.[16] Increasing the concentration of finasteride in the topical solution to 0.25 percent in a later study did alter dihydrotestosterone levels, so care must be shown if and when a topical finasteride is brought to market in picking the appropriate concentration.[17] Why the "if"? The problem with topical finasteride would be an exponentially increased exposure risk to women who may become pregnant and thus impact their offspring. This is likely why we will never see a topical form on pharmacy shelves despite the possibility of eschewing the side effects of oral finasteride.

Finasteride and the Olympics

In a U.S. National Cancer Institute study, 19,000 men ages fifty-five and older who took 5 milligrams of finasteride daily for benign prostatic hyperplasia over seven years showed remarkable results. These study participants experienced a substantial decline in prostate cancer on the order of 25 percent. The study is marred, however, by the abnormally increased rates of high-grade prostate cancer seen in the test subjects who did develop prostate cancer; however, this is thought to be a function of cancer being easier to diagnose on biopsy once the prostate is no longer enlarged and the increased sensitivity of a cancer screening test that interacted with finasteride.[18]

In these last two chapters, it is interesting to see how two medications, finasteride and minoxidil, both achieve the same end goal, one thought to be folly just decades prior. In doing so, however, they bring about hair growth in completely different ways and through different actions on the body, all the while culminating in the production of very different side effects. Not everything is so serious about finasteride, however, as it even rears its head in the world of sports. Why? It is a forbidden substance in Olympic competition and resides on the World Anti-Doping Agency's banned list. Finasteride can act as a masking agent for the steroid nandrolone, with the banned substance finasteride coming to the forefront when Montreal Canadiens star goaltender José Théodore tested positive for finasteride in routine pre-Olympic drug testing in 2006. Théodore, who won the 2002 Hart Memorial Trophy as the National Hockey League's most valuable player—a rare honor for a goaltender—exhibited a noticeably full head of hair yet announced he had been taking finasteride on the advice of a physician for eight years. Any increased muscle mass gained through the use of nandrolone would likely be a hindrance for an acrobatic goalie like Théodore, making it likely Théodore was truly taking finasteride for its intended purpose—glorious hair.[19]

CHAPTER 15

Sildenafil

The structure of sildenafil, the molecule behind the little blue pill. *Structure generated by author in ChemDoodle v 11.5.0, using information supplied by the National Institutes of Health*

It was the end of the 1990s and the twilight of not only a century, but also a millennium. Amidst an impeachment trial and an exploding stock market, what was the First World going crazy about? Sildenafil, the medication better known as Viagra, a little pill making something out of the question for many—an erection—now possible for millions, for a price. Sildenafil is not a cure for erectile dysfunction but a treatment, one that is wildly successful and that has made the drug company Pfizer billions of dollars in the process. Sildenafil, as we will find out, was not an overnight success; the drug was almost abandoned due to its initial failures. Thankfully, its second life was found and with it an amelioration of a condition humankind has searched for throughout nearly all of recorded history.

While sildenafil is the active ingredient in Viagra, the molecular composition present is actually sildenafil citrate. A small citrate molecule sits adjacent to sildenafil to increase overall chemical stability. For the purpose of this chapter, sildenafil will be used to mean sildenafil citrate to simplify things.

Historical Attempts to Ward Off Impotence

Humankind has literally been searching for a treatment for impotence—the inability to achieve or maintain an erection—for thousands of years. One can argue for its appearance in the Old Testament by allusion in the story of Abraham, with there being definitive evidence pointing to the presence of impotence as a known problem on Egyptian tombs (imagine being the gent forever described as impotent at his place of eternal rest—talk about proper reason for a haunting).[1] Frenchmen in the fifteenth and sixteenth centuries could be subject to impotence trials should they be unable to perform in the bedroom, and for a considerable span of time, this was the only way to garner a divorce in France.[2]

As one can imagine, historical treatments and cure-alls for impotence ran the gamut. Greeks and Romans wore talismans of rooster and goat genitalia to ward it off. They also ate animals thought to have high libidos, like rabbits, and drank the semen of hawks and eagles. Such unusual treatments continued for more than 1,000 years, with Albert Magnus, a friar in the 1200s, suggesting that eating roasted wolf penis would cure the issue.[3]

Regardless of the effort and ingenuity put into these "treatments" for impotence, the problem persisted. King Louis XVI of France experienced it as early as age fifteen, but he was far from alone. Even with all of the advancements in modern medicine and nutrition, a majority of men experience this problem at some time in their lives. The 2004 Massachusetts Male Aging Study showed that 52 percent of men aged forty to seventy experienced impotence, but age was not the only factor: underlying cardiovascular disease played a role as well by impairing blood flow.[4]

As time passed and the scientific understanding underlying medicine grew, techniques rapidly changed to alter how medicine was combating the issue of impotence. An early pioneer in impotence treatment, Regnier de Graaf, showed the world in 1668 that he could cause an erection. How did he do it? By injecting saline into the penile blood vessels of cadavers, displaying the power the blood vessels of the region have over the desired event.[5] Later in that same century, Dr. William Hammond used electricity in an attempt to solve the problem of impotence, placing electrodes on the spine, testicles, and penis in what even Hammond considered a displeasing situation.[6] This would be far from the only use of electricity in hopes of getting a response out of a penis, with an 1863 study showing it was possible to induce an erection—albeit in dogs—with electricity.[7]

Charles Edouard Brown-Séquard would stand out in the mid-1860s thanks to his efforts to posit theories correlating changes in the human body as we age and a decrease in the ability to get an erection. Brown-Séquard theorized that injections of semen into the blood of older men would heighten their virility and mental performance. The Frenchman was also interested in the possibility of using hormonal concoctions to increase the amount of testosterone in the body. The French physician injected himself ten times at the ripe old age of seventy-two in 1889 with extracts he made from the testicles of dogs and guinea pigs.[8] Brown-Séquard reported such detailed results as a newly rediscovered "jet of urine" as well as an increase in the "power of defecation." Unfortunately, these enhancements lasted only about a month and were likely the result of a placebo effect.[9] In the early 1920s, a surgical approach reared its head. Russian Serge Voronoff tackled the testosterone problem suggested by Brown-Séquard and took it one step farther, carrying out procedures wherein he would graft the testicles of apes onto the testes of live human patients in hopes of increasing testosterone output.[10] The surgeries were bizarrely successful in the short term, with the increase in testosterone lasting for one to two years before the grafted tissue became fibrotic.[11]

As time passed, more advanced techniques came into play, as did prostheses and "marital aids." The first vacuum device was conceived hundreds of years ago, as it was theorized as early as 1694. A working model of such a device was constructed by Vincent Marie Mondat at the beginning of the nineteenth century and not patented until much later in 1913.[12] A vacuum apparatus works by applying negative pressure to the penis, leading to an increase in blood flow to the organ and an erection within two to three minutes. The erection is then prolonged by the placement of a ring around the base of the penis to prohibit the blood from flowing out and back into regular circulation. This constriction can be an issue, however, as the penis may turn a bluish hue over time when using a vacuum pump device.[13] Despite these side effects, vacuum pumps continued to be popular, even with the longtime prerequisite requirements of a prescription for those in the United States to obtain one. One popular brand was made by Pentecostal preacher Geddings Osbon, who marketed his own version of the pump as the Osbon ErecAid.[14] Osbon labeled the ErecAid as a "youth equivalency device," a name foreshadowing the issue of impotence as one ages.[15]

In addition to the vacuum device, modern medicine introduced a number of surgical implants for those having trouble maintaining an erection. The first well-documented implantation occurred in 1936, when Nicolai Bogoras fashioned an implant from rib cartilage and bone to produce a rigid, erect penis in a patient. Within months, however, the crude implant was reabsorbed by the body, and the procedure was rendered moot. This did not preclude others from broaching the possibility of implantation, with rigid rods of polyethylene showing success in some patients albeit giving the appearance of permanent erection. In due time, silicone rubber would be used to solve this problem, which made for a much more natural surgical outcome. Why? Since silicone rubber is flexible and not likely to be reabsorbed by the body, a small bulb of fluid could also be implanted as well as a pump in the scrotum. This pump and fluid system allowed the individual to send fluid into the silicone implants within the penis to achieve an erection when necessary. Needless to say, this surgical process was quite invasive, and any infection that might arise in the years following would call for the removal of the entire prosthesis and pump.[16] The alternative to an implanted silicone rubber prosthesis and pump was the cleverly designed silver metal and silicone penis prosthesis, which consisted of two rods of silicone and a silver wire core that could be used to flex and bend the penis into the desired orientation.[17]

If one did not want to go the route of an implant, one could use the direct injection of substances into the penis to achieve an erection in what is known as intracavernosal injection therapy. Intracavernosal therapy consists of the injection of a medication, often alprostadil, directly into the penis, with the subsequent relaxation of the smooth muscle tissue of the penis leading to an erection within minutes. Despite the fear that might set in while injecting oneself in the penis prior to intercourse, this was quite a popular therapy thanks to its being successful in nearly 80 percent of patients.[18] Intracavernosal therapy mingled with undesired showmanship during the 1983 Urodynamics Society Conference in Las Vegas thanks to Sir Giles Brindley's awkward and unforgettable (and even emotionally scarring) lecture. A number of attendees noticed Brindley in track pants while boarding the elevator just moments prior to his appointed lecture time. In the following hour, he proceeded to give a lecture revolving around the discoveries he made while injecting his own penis with the vasoactive drugs phentolamine and papaverine, complete with slides of his handiwork. Nearing the end of his lecture and before the eighty or so people in attendance, Brindley announced he had injected himself with papaverine in his hotel room moments before the lecture started, and deemed it reasonable to pull his track pants back in such a way as to show off his erection. This, for some reason, was not enough for Brindley, who then decided to pull off his pants and wade into the audience so that the evidence of the injections could be made fully known. Shouts and screams accompanied this move, and before long, Brindley was brought back to reality, sheepishly realizing what he had done.[19] Despite Brindley's unfortunate showing, the entire problem of erectile dysfunction would be revolutionized fifteen years after his memorable lecture.

Sildenafil's Early Failure

Sildenafil was initially developed by Pfizer as molecule UK-92,480 by Peter Dunn and Albert Wood in their laboratories in Sandwich, Kent, a small town in southeastern England. Dunn and Wood were looking for a molecule that would interact with the enzyme phosphodiesterase type 5 in hopes of treating high blood pressure and angina, a pain in the chest due to a reduction in the flow of blood to the heart. Pfizer's Nicholas Terret is also named as one of the inventors of UK-92,480, receiving an inclusion on the patent itself.[20] UK-92,480 is a rather bland name for the molecule that would one day become Viagra, but it represents the nomenclature of location—the U.K. Pfizer laboratories—and either the notebook page or volume number or its place in the sequence of molecules synthesized at the location.

The initial studies of UK-92,480's effect on angina and blood pressure were by all accounts disastrous. The 475-dalton molecule should be performing; it worked well in animal studies at least. By June 1993, Pfizer senior management saw the sildenafil project as a money pit, and if the results did not fare better within three months, the project would be shelved. Hope broke through in a nearby set of trials on miners in South Wales wherein during the debriefing session, a number of the participants reported experiencing more erections than normal.[21] There are also reports of men refusing to give back their remaining medication at the end of the trial.[22] This would correlate with reports of nurses observing that a lot of men were found lying on their stomachs when they came to check in on them during the initial studies. The gentlemen were doing this to hide their unexpected erections, showing that sildenafil did work to dilate blood vessels, just different ones than planned—blood vessels in the penis and not the heart.[23]

Excited by this unexpected side effect of UK-92,480, Pfizer chemist David Brown immediately went to Pfizer Research and Development asking for £150,000—a paltry sum in pharmaceutical circles—to perform an impotence study using UK-92,480. Brown refused to leave the office of his superiors until he secured the funds. Soon afterward, a study started in the city of Bristol before expanding to larger European sites.[24] These clinical studies obtained numerical data from a device named the RigiScan, which was fitted on the trial participants as they watched adult movies.[25] The results were better than anyone could have imagined. Sildenafil turned out to be practically the perfect drug for erectile dysfunction. It acts within thirty to sixty minutes, a short enough time that a patient can take the pill when they are reasonably sure a sexual encounter will occur. It's a pill, and the fact that it is so aids in Viagra being discrete and easy to take—one can swiftly pop a pill with none the wiser instead of sildenafil's harbingers, which required a priming of the genitals with a pump or direct injections. Viagra also has a short half-life, with most of the drug falling out of use within four hours, not lingering on for hours after any sexual encounters have ended. It also leaves no appreciable increase in heart rate or blood pressure, a key aspect, as many of those taking sildenafil would likely exhibit a heart condition that predisposed them to erectile dysfunction in the first place.[26]

The ability to overcome erectile dysfunction by simply popping a pill was such a sea change and far better than injections directly into the penis or surgery. It also took the treatment (important to note, as sildenafil is still just a treatment for erectile dysfunction, not a cure) of impotence out of the urologist's office and placed it within the realm of the general practicing physician, who could easily prescribe such a pill with a proven track record and easy-to-follow instructions.

Sildenafil also benefits from a readily understood mechanism of action. Sildenafil works to promote erections by negatively interacting with phospho-diesterase type 5 (PDE5), the same enzyme the original researchers of Viagra in Sandwich, Kent, were targeting to inhibit to combat angina and high blood pressure. In binding to PDE5, sildenafil turns off the enzyme and prevents molecules of cyclic guanosine monophosphate (cGMP) from being modified and changed into their noncyclic form, GMP. Before this occurs, however, we need to back up a few steps to find out why cGMP is present in the first place. cGMP is increasingly present when nitric oxide (not to be confused with nitrous oxide) is released by the nerves along the penis due to sexual stimulation and proliferates throughout the smooth muscle of the organ. This leads to a relaxation of the muscle, allowing blood to flow into the corpus cavernosum, and, eventually, an erection. If there is no stimulation, there is no cGMP for PDE5 to interact with (or not interact with in the presence of sildenafil) and no erection. Normally, cGMP would be altered by PDE5, but molecules of sildenafil inhibit PDE5, allowing the cGMP to stay around longer and signal for the relaxation of smooth muscle tissue and the increased flow of blood to occur. Once again, it's important to note that without the release of nitric oxide, there is not sufficient cGMP present for an erection, meaning sildenafil has no effect unless sexual stimulation also takes place.[27]

In 1996, Pfizer patented sildenafil citrate in the United States.[28] Sildenafil soon became Viagra, with the name formed from "vi-" for "vitality" and "-agra," suggesting an air of accomplishment through the cultivation of pharmaceutical ingenuity amidst a series of setbacks.[29] In less than two years, Pfizer would take sildenafil citrate from patent to market, a lightning-fast approval time, as the U.S. Food and Drug Administration (FDA) approved Viagra for use to stem erectile dysfunction on March 27, 1998.

Impotence Becomes Erectile Dysfunction, Courtesy of Pfizer Marketing

But before sildenafil could reap Pfizer a financial windfall, Pfizer artfully altered the world's perception of impotence in hopes of attracting more men to take the drug. A large part of this was rebranding the problem at hand—impotence—as erectile dysfunction in their literature and media advertisements to reduce social stigma.[30] Marketing also surmised that more than half of men over age forty likely suffer from erectile dysfunction. This was a positive conclusion for their sake, as it made erectile dysfunction more common and by extension made it a problem to be treated and not a source of shame.[31]

The first commercial was a timing boon for Pfizer as direct-to-consumer advertisement regulations had just been eased by the FDA. This first spot starred former presidential candidate Bob Dole, speaking about erectile dysfunction amidst the backdrop of the Bill Clinton–Monica Lewinsky scandal. Earlier in 1998 and prior to Bob Dole's official role as Viagra spokesman, the former presidential candidate spoke on *Larry King Live* about his own problems with sexual dysfunction, stemming from a prolonged battle with prostate cancer.[32] Initially, U.S. television decency restrictions forbade the advertisement be shown before 11 p.m. despite the fact that the first commercial was rather tame. In the commercial, Dole toes a courageous line, speaking of his battle with prostate cancer and subsequent erectile dysfunction, with Pfizer using the opportunity to deftly rebrand erectile dysfunction as "ED." Dole never once goes into any details about what erectile dysfunction is or the availability of a new drug, Viagra, to treat it, instead asking viewers suffering from "ED" to get a check-up.[33]

Over the years, Pfizer's commercial approach would evolve. As a part of Pfizer's Viagra push, the company sponsored the number 6 car beginning with the 2001 NASCAR Sprint Cup Series, with Hall of Famer Mark Martin driving the car for five seasons. The forty-two-year-old Martin was the perfect age to represent a potential Viagra user in 2001, and the emblazoned car did much to promote the drug and decrease stigma despite going down in the eyes of some as the most embarrassing NASCAR sponsorship deal in history. I politely disagree, and would like to point to the 2004–2005 sponsorship of a car by Boudreaux's Butt Paste—as one who has used the cream many a time on his infants—as the most embarrassing NASCAR sponsor of all time.[34] Pfizer also obtained the services of Rafael Palmeiro, star of Major League Baseball's Texas Rangers and Baltimore Orioles, for a cool $2 million in 2002 to star in a commercial. Palmeiro broke from the Bob Dole mold in the thirty-second spot wherein the then thirty-eight-year-old slugger utters the words, "I take batting practice, I take infield practice, I take Viagra," in the boldest declaration of use by a spokesperson to date.[35]

Along with the advertising rollout came a bevy of other changes, including sexual harassment training for Pfizer sales representatives, both male and female, who now would be talking about erections in the workplace for hours on end.[36] Pfizer even sent a delegation to the Vatican to investigate its view of the new medication.[37] Along with the ethical quandaries, the arrival of Viagra also brought on an interesting question for insurance companies, such as, "Should we cover such a drug for those in their sixties?"[38] Stories traveled of patients coming to appointments in trench coats, hats, and sunglasses, refusing to give their real name in hopes of keeping some shred of perceived pride through anonymity. Time would change this perception, with some physicians' offices altering schedules to become weekend Viagra hot spots with appointments scheduled every ten minutes to meet patient demand.[39] The Viagra effect was not only an increase in erections; it also brought more men into physicians' offices and led to a net good, as many began being treated for the underlying systemic medical issue that led to their erectile dysfunction.

Despite the good-natured humor and overwhelmingly positive reception of Viagra, questions remained. In time, more studies, especially long-term studies featuring men regularly taking Viagra, would be carried out to determine sildenafil's safety. A four-year study of nearly 1,000 men taking sildenafil approximately twice a week showed no long-term issues with regular use of the medication. The average age of the men involved was fifty-eight, an intelligent population build, as it was old enough to predispose a larger percentage of the study population to heart conditions and further exhibit sildenafil's safety. The study was not without incident, as during its course, eighteen heart attacks were reported. Further examination, however, showed sildenafil to not be the underlying cause of the myocardial infarctions. The study participants were also given the choice of twenty-five-, fifty-, or 100-milligram Viagra for their use, with nearly 90 percent of those in the study going with the maximum dose. This study did wonders for dispelling rumors and showing the safety of the medication in a population making frequent use of the drug to fight erectile dysfunction.[40]

The year 2003 would see two challengers to Viagra's throne arise in the form of two new erectile dysfunction drugs—Bayer Corporation's Levitra (vardenafil hydrochloride) and Lilly USA's Cialis (tadalafil).[41] Cialis is a very different molecule from Viagra. Levitra, however, closely resembles Viagra, at least visually, as the only difference between the two molecules is that Levitra exhibits an alteration in the placement of a nitrogen atom and the addition of a methyl group to one end. Both Cialis and Levitra are potent PDE5 inhibitors, performing in much the same way as Viagra.

The Very Visual Side Effects of Viagra

The very drug touting the ability to treat erectile dysfunction also had some very embarrassing side effects, foremost of which is priapism. One suffering from priapism has the telltale sign of an erection lasting for longer than four hours. In reality, priapism is not a laughing matter, even if the medical malady itself is named after the minor Greek fertility god Priapus, who was often shown as having a humungous permanent erection in paintings and frescoes. If left untreated, priapism can result in tissue damage or, in a circular turn of events, erectile dysfunction. Priapism can also develop in those living with sickle cell anemia, as the sickle-shaped red blood cells block the return flow of blood out of the penis.

A tinge of blue in one's eyesight is another side effect of long-term sildenafil citrate use. This side effect, cyanopsia, occurs due to sildenafil citrate acting on a different phosphodiesterase than PDE5. In cases of cyanopsia, the drug acts on PDE6 as well as PDE5, but it does so to a lesser extent. PDE6 regulates the sensitivity of the rods in your eye, and the inhibition of PDE6—just as sildenafil inhibits PDE5—in the presence of sildenafil leads to a hypersensitivity of the rods and an increase in blue in one's field of vision.[42] Taking a large amount of Viagra at one time is also cause for concern. Vision problems due to a damaging of the outer retina, including a general blurriness and color distortion along with dizziness and a flushed appearance of the skin, were reported in a Beijing female when she ingested 2,000 milligrams of Viagra "impulsively" (as the case report states) after an argument with her spouse. This amount is the same as twenty of the high-dose 100-milligram pills, with it very unlikely that anyone would take this much sildenafil citrate.[43] Light sensitivity and red–green color blindness that lasts for longer than twenty-four hours have also been associated with use of Viagra.[44]

Counterfeit Pills and Illicit Use

In the first decade of the drug, more than 30 million prescriptions for sildenafil would be filled. This enormous number still did not meet the demand for Pfizer's little blue pill, as drug enforcement offices often came across counterfeit Viagra containing plaster of Paris, dust from bricks, and arsenic. Clandestine uses for sildenafil began to pop up, with younger men free from the typical form of age and cardiovascular-driven erectile dysfunction taking the drug to counter the effects of recreational drugs and alcohol that were eliminating their ability to achieve an erection.[45]

In a study of sixty "young" men between the ages of twenty and forty without erectile dysfunction, researchers aimed to determine if sildenafil played any role in their sexual performance. One group of study participants would take twenty-five milligrams of sildenafil in the minutes leading up to sexual intercourse, while the other group of participants would be given a placebo. What the study found was quite interesting. Although there was no change in the quality of erections of these healthy men, what did change was a decrease in refractory time (the time from orgasm to the moment one could become sexually aroused once more). In the group taking sildenafil, the refractory time dropped nearly ten minutes, from 14.9 minutes to 5.5 minutes, a substantial decrease. Interestingly, the placebo group also showed a ninety-second decrease in refractory time, along with a statistically insignificant number of placebo patients claiming it was easier to achieve erections. While this shrinking of the refractory period is not what sildenafil is pronounced to do, it could be useful for those suffering from consistent premature ejaculation problems, which does not have a bevy of treatments at the moment and lacks a cure.[46]

Pfizer, on the back of Viagra sales, rose to become the fifth most profitable corporation in the United States in 2002. In the early days of the Viagra craze, the individual little blue pills came in at $8 to $10 each at your local pharmacy[47]— pricey on the one hand but a small price to pay for those looking to reopen a chapter of their lives they thought had evaporated long ago. Viagra also took a burden off of women in the bedroom, who often, quite unfairly, played the culprit for decades when it came to their male partner's inability to perform sexually.[48] Pfizer's U.S. patent on sildenafil fully expired in 2020, paving the way for millions to get the drug in a generic form at a lower cost.[49] I say "fully" here, as Pfizer reached an agreement with the generic pharmaceuticals giant Teva allowing Teva to produce a generic form three years prior to patent expiration.[50]

In the past years, a cottage industry of pharmaceutical firms has sprung up offering alternate formulations of sildenafil citrate, moving from the pill to sprays, chewables, and under-the-tongue dissolvable lozenges. These companies also help the end user by providing an electronic means to obtain a prescription via an online consultation with a physician. In the process, the companies are eliminating another hurdle to getting the prescription drug in the United States as well as giving an added air of anonymity. In the United Kingdom, it's even easier to obtain sildenafil citrate: it is an over-the-counter medication. Pfizer teased an application for over-the-counter sales of Viagra to the European Medicines Agency in 2008 as an erectile dysfunction treatment, but Pfizer eventually abandoned the endeavor, as it felt the medication needed a physician's supervision. This started the push for over-the-counter Viagra, with the U.K.-based pharmacy chain Boots opening up to sales of Viagra without a prescription in 2009, with the Tesco chain following suit in 2010.[51] The over-the-counter push did not extend to the United States, as the medication, generic or otherwise, continues to require a prescription.

There has not been a lot of success in finding a medication that works in females similar to how Viagra works in males. The closest would be the drug Addyi, which boosts blood flow to the pelvic region and raises levels of serotonin and dopamine. Addyi has been available in the United States since 2015 but carries with it side effects of sleepiness and sudden drops of blood pressure leading to fainting, preventing its wide-scale use.[52]

Viagra Plays a Role in U.S. Foreign Policy and Later Gets a Second Life as Revatio

Viagra has also seen some more unusual uses. SENA, the state culinary school of Colombia, got on the Viagra publicity train (albeit a few years down the road), creating a dessert for a 2009 event combining passion fruit, chocolate, whipped cream, and an unknown quantity of sildenafil into the pudding. The school lacked the ability to sell the dessert at the time, as it would have to receive approval from the Columbian Food and Drug Institute, but the dish made for an interesting proof of concept nonetheless.[53]

There are reports that the U.S. Central Intelligence Agency (CIA) used Viagra as a bargaining tool with informants in Afghanistan to gain a one-up on the Taliban in the region.[54] The CIA was experiencing trouble gaining access to the village of a sixty-year-old South Afghanistan chieftain, one who was particularly wary of U.S. intervention. Along with access to the key passages he controlled, they needed his knowledge base. Aware of his age and its particular disposition to erectile dysfunction, a field operative gave the clan leader four tiny Pfizer logo–emblazoned sildenafil citrate tablets and by the next day won his loyalty.[55]

Sildenafil did, eventually, get approved to treat a heart condition in the fashion Pfizer's Wood and Dunn pined for in the late 1980s and early 1990s. Pfizer gained FDA approval in 2005 for sildenafil's use to fight pulmonary arterial hypertension after a new series of clinical trials. Pulmonary arterial hypertension is a disease denoted by narrowing blood vessels in and around the lungs and is characterized by high blood pressure in the arteries delivering freshly oxygenated blood to the lungs. With this approval, sildenafil began being sold in a lower-dose form than Viagra under the brand name Revatio.[56]

Viagra may yet still have more uses, including helping travelers, both male and female, recover from jet lag. Hamsters given a small amount of sildenafil acclimated to a day/night cycle almost twice as fast as their fellow hamsters given placebo. The hamsters receiving sildenafil overcame simulated jet lag through a change in their light and dark cycles in just eight days compared to twelve. When the amount of sildenafil was increased by the scientists undertaking this bizarre study, the hamsters receiving sildenafil could overcome jet lag in as little as six days, but it came with a price—hamster erections. The study involved only male hamsters at the time, but future studies aimed to include females in hopes of finding a cure for jet lag in us, their fellow mammals.[57]

Hamsters, hearts, and erections. The application of sildenafil truly runs the gamut. Not bad for a medication almost given up on by corporate executives yet saved from the trash heap by scientists willing to look for results in unusual places and act on them.

CHAPTER 16

Quirks of Pharmacology and the Pharmaceutical Industry

You have no doubt put up with a lot of quirkiness so far in this book (and hopefully it was not all bad), but for a variety of reasons, many aspects of pharmacology and the pharmaceutical industry are shrouded in a veil of mystery. Why are you unable to buy an antibiotic over the counter in the United States to cure your child's strep throat? Ever wonder why diabetics cannot drink insulin and instead have to rely on pumps or injections or why drug companies keep advertising directly to you, even for medications you cannot go to your local pharmacy and buy without a prescription? Let's tackle these questions and a few more as we pull back the curtain and take a deep dive into some quirks of the pharmaceutical world.

If You Have Diabetes, Why Can't You Drink Your Insulin?

Insulin is not like most other pharmaceuticals, which are often small molecules. It is a peptide hormone created in the pancreas and a large one at that, weighing in at just over 5,800 daltons, or about the same as 484 atoms of carbon. That is pretty darn big. Insulin is present in your body to help it maintain proper levels of glucose (a sugar commonly broken down from carbohydrates) in your blood. To give another comparison of insulin's massive size, glucose is only 180 daltons. Insulin is released by your body in response to increased blood glucose levels, essentially telling your body it is time to make use of energy sources at the cellular level.

The discovery of insulin came in stages, stemming from a medical student by the name of Paul Langerhans. Langerhans observed piles of cells on a pancreas in 1869.[1] These little collections of cells would later become known as the islets of Langerhans, with many canines becoming unwary test subjects over the next several decades as the connection between diabetes and the pancreas was established and put to rest by the 1920s.

Most insulin sold in the world today is biosynthetic "human" insulin. This type of insulin is made by leveraging technology wherein human DNA is inserted into *Escherichia coli* cells and the insulin harvested from cell batches at the proper time. Prior to this genetic approach, the majority of the world's insulin came from cows and pigs, as cow and pig insulins are nearly identical in structure to human insulin. Two tons of pig organs were needed to acquire eight ounces of insulin.[2]

Insulin is traditionally administered subcutaneously through a series of injections during the day via a reasonably small handheld pump, though this was not always the case. The first insulin pump was created in 1963 and was roughly the size of a backpack. But why go through all that trouble with injections— why can't diabetics just drink insulin? It's because it tastes bad: it is made from two tons of pig organs. I'm kidding. At the structural level, insulin is a series of fifty-one amino acids. Amino acids are what proteins are made from, and as you probably know, proteins are broken down by your body to be shunted for use throughout. Your body is keenly attenuated to break apart proteins. If the fifty-one amino acids in insulin ever reached your gastrointestinal tract, the insulin would be torn apart into tiny fragments well before they could start and participate in any blood glucose–lowering cascade that would bring relief to a diabetic. By injecting directly into the thigh or abdomen, one bypasses the gastrointestinal tract altogether, allowing the insulin to signal the body correctly.

How Do Extended-Release Pharmaceuticals Work?

Extended-release drugs are those that deliver their drug payload either steadily over a given period (controlled release) or not so steadily yet still during a period of time (sustained release). Extended-release drugs aim to make taking a medication easier. If fewer doses of a drug are necessary during the day, patient compliance increases, all the while maintaining what is hoped to be a steadier concentration of the drug throughout the day than seen with an instantaneous-release form of the same drug. Most extended-release pharmaceuticals are created by surrounding a core of the active ingredient in the medication with layers of polymer. This yields a polymer maze that the active ingredient must find its way out of like an aimless spelunker through the process of diffusion once the pill is swallowed. These polymer exteriors can also be surrounded by a gel-like substance that swells and further slows the exit of the active ingredient.

Slightly more elegant methods of extended-release formulations also exist. One of the more interesting techniques is known as the osmotic controlled-release oral delivery system (OROS). In an OROS medication, a tablet containing the active ingredient is coated with a covering containing one or more laser-drilled holes. Once an OROS medication is swallowed, it takes in water through the laser-drilled openings. As water from your body enters the OROS tablet, the water rushes in and pushes out the active ingredient in the tablet over the course of this osmotic interaction and maintains a very reliable form of extended release.[3]

It may go without saying, but extended-release medications are not to be split or crushed, for, as you can tell by the methods above, this would hamper their modified release mechanism. Which type of extended-release method is used by a pharmaceutical company varies from pharmaceutical to pharmaceutical, with consideration taken into its effect on cost and, most importantly, maintaining a proper concentration of the drug over the delivery period. Extended-release forms of medications whose patents are about to expire are also popular for pharmaceutical companies, as they give a second chance to recoup money from a medication they are about to lose total control of as it becomes a generic.

Passing through the Blood–Brain Barrier

The blood–brain barrier poses a major conundrum for any researcher hoping to introduce a molecule into the body whose goal is to interact with the brain. The barrier is formed from endothelial cells, the same cells that line blood vessels, and is present to keep toxins away from your precious neurons.[4] Think of it sort of like a checkpoint for your brain—it will let through water and glucose (much-needed energy) in the blood and other resources, but it keeps pathogens like viruses and bacteria that might inhabit your bloodstream out of your brain and cerebrospinal fluid. A lucky number of pharmaceuticals are able to pass through the blood–brain barrier unaided (caffeine can as well) and act on the brain. However, for most drugs to be that are hoping to act on the brain, the blood–brain barrier becomes the stopping point in their development.

The blood–brain barrier accomplishes its job as the brain's bouncer through a series of tight junctions spread among the endothelial cells. These tight junctions are formed by transmembrane proteins constantly on guard against large molecules (unfortunately, some drug candidates fit this bill) and possible invaders looking to come in contact with the brain. It's protection, pure and simple. If this protection wanes, the results can be dire. "Leaking" of the blood–brain barrier hastens the onset of a variety of neurological diseases, including Alzheimer's and multiple sclerosis.[5]

Any pharmaceutical hoping to act on the central nervous system must be able to pass through the blood–brain barrier. All the work the blood–brain barrier is doing renders a slew of possible neuropharmaceuticals useless. How do researchers trick the blood–brain barrier into allowing their particular molecule through so it can act on the brain? By creating molecules that can pass through the endothelial cells and by staying away from the tight junctions altogether. If a molecule is able to pass through the cell membrane of an endothelial cell, it might pass through the blood–brain barrier. Even once it is past the blood–brain barrier, a number of maladies can occur, as the drug may not get through in high enough concentration. The medication could also be modified en route by the body and its effect made null.

Drug designers are constantly looking for solutions to this problem, with two major ones standing out. One is the use of "pro-drugs." The pro-drugs concept involves taking a drug molecule shown to work in the laboratory and attaching more and more atoms to it to increase its lipophilicity and thus its ability to mimic the membrane of endothelial cells, then hope that the non–medically active portion is sheered away, in this case, en route to the brain. With a pro-drug, you are designing not the version of the molecule that is active in the body but rather a version that will survive in the body long enough to become active. In the second, the use of nanoparticles, the drug molecule is attached to a nanoparticle that is known to be able to pass through the blood–brain barrier. In one case, this is made possible using human serum albumin, which is already present in your blood and blood plasma. By attaching your drug molecule of choice to human serum albumin, the molecule sneaks by the blood–brain barrier thanks to the ability of the human serum albumin to freely pass through the blood–brain barrier, as it does not identify the albumin as a foreign entity.[6]

Other tricks for passing the blood–brain barrier are in the works, including using high-intensity focused ultrasound to disrupt the blood–brain barrier long enough for a dose to be administered, but we will not be seeing that coming to us in a pill form any time soon.

If One Pill Is Good, Why Are Two Pills Not Twice as Good?

Your head hurts. I mean, it really hurts. The packaging for your analgesic of choice says to take one pill every four hours, but man, does your head hurt. Can't you save yourself some trouble and get rid of the pain quicker by taking two or even three? No, not necessarily.

The window between therapeutic application and toxic exposure is sometimes small. This is often (and thankfully) not the case with over-the-counter pharmaceuticals, but with prescription drugs, it can be a thin line and open the patient up to additional side effects or a dire situation.

To determine the proper dosage, or why you need just one pill at a time for our hypothetical headache medication, researchers measure the blood serum concentration of the drug over time to determine the effective dose. When taking a medication, once the effective dose (the desired concentration of drug in your bloodstream) reaches a certain point, taking any more often fails to produce a positive effect. Imbibing any additional medication may in fact open up the patient to additional side effects as they wander outside the therapeutic window of the drug. Why? Because almost all the desired receptors able to bind our hypothetical headache medication are bound by the drug, allowing additional molecules of the drug to be taken up by promiscuous receptors (a great possible band name by the way) that will bind to a variety of small molecules. This unwanted binding event yields no further benefit and in many cases generates unwanted side effects.

How Does a Prescription Pharmaceutical Become an Over-the-Counter Medication?

Over-the-counter medications are pharmaceuticals you can buy without a physician's prescription, thus allowing a patient to "self-treat" a condition. These medications can be a lifesaver when a patient is unable to schedule an appointment with a physician but is able to understand the source of their pain or discomfort. In the United States, over-the-counter medications are required to carry an OTC Drug Facts Label detailing how the drug should be taken, its active ingredients, its use, specific warnings, how often and how it should be dosed, and any inactive ingredients. The labels are written with older populations in mind, as this is the subset of society that purchases nearly 30 percent of the over-the-counter medications sold in the United States.[7]

It is also possible for a formerly prescription drug to become an over-the-counter offering. For example, the popular over-the-counter medications Rogaine, Flonase, and Allegra all started out as prescription pharmaceuticals. At the core of transitioning a prescription pharmaceutical to over-the-counter access is the patient's ability to easily self-diagnose their malady. This is overcome with some problems, as one can self-diagnose with some reliability a headache, male-pattern baldness, or seasonal allergies. To undergo a switch from prescription to over the counter in the United States requires FDA approval that the medication is safe and unlikely to produce severe side effects if the patient self-medicates incorrectly. Also, if the prescription drug has an addictive quality, it is ineligible in most cases to become an over-the-counter drug.[8]

Restricted over-the-counter medications also exist. These "behind-the-counter" pharmaceuticals often require an additional hurdle for the consumer to purchase. For example, to purchase pseudoephedrine in much of the United States requires the presentation of a driver's license or another form of state identification. This serves to monitor the amount of pseudoephedrine purchased by an individual and ensure it is under predetermined monthly limits. Pseudo-ephedrine is subject to abuse on its own or for use as a precursor to methamphet-amine hydrochloride, better known as the illicit drug crystal meth.

Why Do Antibiotics Require a Prescription in Some Countries?

In the United States, you cannot walk into a pharmacy, cash in hand, and pur-chase a typical oral antibiotic. Topical antibiotic creams for cuts, yes, but ones to treat strep throat, for example, no. Antibiotics do not exhibit any addictive qualities, and they are difficult to overdose on, so why are they not available for over-the-counter purchase? The answer comes down to concerns over antibiotic resistance and whether a strain of bacteria is able to gain a permanent foothold in your personal flora and fauna. If antibiotics are taken incorrectly—either for too short of a time period or at too low of a dose—the symptoms of your infec-tion might go away but not all of the bacteria causing the infection. Leftover bacteria can then gain resistance to the antibiotic you used through successive generations of proliferating, and the next time you have an infection or abscessed tooth and took the same antibiotic, the bacteria would be unphased. This prob-lem would only be compounded if multiple antibiotics were made available over the counter. This could lead to "antibiotic shopping" as you move from one to another in hopes of finding one that works. Antibiotic shopping would open an individual up to increasing resistance issues and limit the benefit from different types of antibiotics should a serious problem requiring hospitalization arise. Placing antibiotics behind a prescription barrier limits access but also limits the opportunities for people to dose themselves into bacterial resistance.

What Are Off-Label Prescriptions?

Off-label prescribing is the prescribed use of a medication for a treatment that the FDA has not approved. It is not illegal, and, in fact, it is a very common way to treat diseases states for which a drug company has not sought full FDA approval due to cost or, in cases where the drug being prescribed off label is available only as a generic, lacks a funding sponsor.

Once the FDA has approved a drug for a use, it is up to the physician to use their judgment based on clinical evidence as to whether the medication would be applicable for an as-yet-unapproved use. Why would a health care provider prescribe off label? They do so in situations where no drug yet exists to treat a disease state or where none of the approved drugs are showing signs of success with the patient. The former is particularly pertinent when a cancer drug approved for one type of cancer is used for another cancer for which no approved chemotherapy exists. Off-label use can extend to dosages and dosage forms as well wherein a drug may be actually approved for the condition, just not in the prescribed amount or form.[9]

Examples of off-label prescriptions are prevalent, such as when the allergy medicine diphenhydramine is prescribed to treat insomnia, when tricyclic antidepressants are prescribed for neuropathic pain, or when atypical antipsychotics like risperidone are prescribed for eating disorders or obsessive-compulsive disorder. Attempts by drug companies to advertise off-label uses are typically frowned on, but pharmaceutical manufacturers can distribute journal articles and textbook chapters supporting an FDA-unapproved use to health care providers to encourage off-label prescribing.[10]

How Do Individuals with Extremely Rare Diseases Get Medications for Their Conditions?

Say you are unlucky enough to become infected with dengue virus or find yourself diagnosed with an extremely rare form of cancer like glioblastoma. What hope is there for you to get the medication you need if there is an extremely small market for your lifesaving pharmaceutical?

The FDA established the Office of Orphan Products Development through the 1983 Orphan Drug Act, with more than 600 so-called orphan drugs researched and developed in the four decades since. A pharmaceutical is designated as an orphan drug if it is intended for the treatment of a disease state from which fewer than 200,000 individuals in the United States suffer.[11] Such treatments, due to a small patient population, would never recoup the hundreds of millions of dollars needed to fund research, development, and manufacturing on the part of a pharmaceutical company let alone turn a profit. Not all of the treatments emanating from the Orphan Drug Act are the typical small-molecule pharmaceutical either, as many are much more expensive biological products (e.g., gene therapy or a complex combination of protein).

What do the companies get for pushing an orphan drug through to market? Under orphan drug status, pharmaceutical companies receive seven years of FDA-mandated exclusivity on the market along with tax incentives in the form of tax credits totaling as much as 50 percent of research and development costs. Funding through grants for research and development are also available for orphan drugs.[12] Such generous measures have given a new lease on life to hundreds of thousands of people over the past four decades.

These generous benefits do not prevent abuse or the finding of loopholes, however. In 2016, AstraZeneca applied for orphan drug status for its $2.8 billion-a-year cholesterol drug Crestor for use in homozygous familial hypercholesterolemia, a disease that afflicts a few hundred children each year. Why would Astra-Zeneca apply for orphan drug status for a proven moneymaker with an abundant patient base? AstraZeneca stood to lose exclusivity of Crestor in 2016, opening it up to generic manufacture. As such, AstraZeneca believed the seven years of FDA market exclusivity that comes with orphan drug status would prevent the sale of any generic forms of Crestor for both high cholesterol and homozygous familial hypercholesterolemia.[13] AstraZeneca lost the argument within weeks and was unable to block generic forms of Crestor from reaching the market.[14]

Drug companies can still profit from an orphan drug, however. Alexion Pharmaceuticals manufactures Soliris, earning exclusive rights to the drug for its use in treating the blood disorder atypical hemolytic uremic syndrome. At the height of its sales, Solaris cost between $500,000 and $700,000 per patient for a year's treatment. Most of the cost was taken on by insurance companies, leading Alexion to a $24 billion market valuation in 2017.[15]

What Is Drug Compounding?

Drug compounding is the alteration of an existing commercial pharmaceutical by a physician or pharmacist through the mixing or addition of ingredients to create a medication form tailored to the specific needs of the patient. For example, if a patient, particularly a young one, is unable to swallow a pill and the drug comes only in a pill form, a licensed pharmacist might carefully create a liquid containing the pill that would still meet the needs of the patient for their given malady. While the medication may be FDA approved, once it is compounded, the final form is no longer FDA approved.

Drug compounding is generally quite safe and carried out in ultraclean conditions, but some horror stories abound. In 2012, a fungal meningitis outbreak stemming from compound drugs sickened nearly 800 patients and killed almost 100.[16] The center that performed the compounding in question specialized in steroid injections. The Centers for Disease Control traced the meningitis outbreak to 17,000 vials of methyl prednisolone acetate contaminated with fungus and shipped across twenty-three states.[17] Methyl prednisolone acetate is a steroid and in this case was distributed as an epidural injection for pain relief. In 2018, the owner of the compounding center and four workers (two clean room pharmacists, a verification pharmacist, and the director of operations) were convicted in federal court for their roles in the contamination and ensuing nationwide meningitis outbreak.[18]

Drug compounding woes reached into the veterinary world in 2019, when three horses died after ingesting an incorrectly compounded paste of toltrazuril and pyrimethamine that contained "18 to 21 times the pyrimethamine indicated by the labeling."[19] The compounded paste was intended to treat equine protozoal myeloencephalitis, a parasitic infection of the horse's central nervous system. This is not the first time this mistake was made. In 2014, high doses of pyrimethamine were found in the same mixture intended to treat equine protozoal myeloencephalitis and resulted in the deaths of four horses.[20]

What Is Pharmacological Half-Life?

How long does a drug stay in a person's body? That is the question at the core of metabolic half-life. Half-life is the measure of how long it takes for a pharmaceutical to go from its highest concentration in your body to half of its highest concentration as your body breaks down the drug through various metabolic processes. The length of the half-life governs dosage and how frequently you need to take the drug. For instance, the half-life of the popular benzodiazepine clonazepam (brand name Klonopin) is a little over a day, meaning the drug stays in your system a reasonable amount of time after taking it, allowing it to be prescribed in smaller doses and for longer time intervals. On the other end of the spectrum are analgesics like ibuprofen and acetaminophen, which exhibit a half-life of closer to two hours, explaining their "every four hours" dosage pattern.[21] Half-life also plays a role in whether a drug would be better suited to be an extended-release candidate. Drugs with shorter half-lives that need to stay in the body throughout the day would fall into this category so that a therapeutic dose of the drug is maintained.

Are Dietary Supplements FDA Approved?

Dietary supplements are not subject to the FDA's New Drug Application approval process, making any professions of effectiveness on the part of the manufacturer null and void in the eyes of the FDA. The FDA regards dietary supplements as foods. As such, the organization is interested only in their explicit labeling as a supplement and their safety in the general population after they reach the market. Among these dietary supplements are your typical vitamins along with herbal supplements, amino acids, and enzymes you might see touted for sale with remarkable claims. The FDA does not require by law these claims to be truthful in the eyes of the FDA but does deem it illegal for dietary supplement to be marketed as the treatment or cure for an existing disease state. The FDA also makes note that taking dietary supplements with prescription and over-the-counter medications opens the individual up to possible side effects and urges individuals to let their health care providers know about any dietary supplements they are taking.[22]

Any new dietary ingredient in a dietary supplement must be brought before the FDA no later than seventy-five days prior to going on sale. The FDA defines "new" rather broadly as any dietary ingredient not sold in the United States prior to October 15, 1994, when the Dietary Supplement Health and Education Act was passed.[23] All that is required for dietary supplements in the premarketing phase is the manufacturer's basis for why they believe the new dietary ingredient to be safe. If a problem with safety stems from the ingredient after it is on the market, however, the FDA steps in to monitor, research, and evaluate the safety of the new dietary ingredient and deem whether the supplement should continue to be for sale.[24]

What Is a Formulary, and How Does It Affect What I Am Prescribed?

Simply put, a formulary is a set of medicines that a hospital, a system of linked hospitals, or an insurance company is willing for its health care providers to prescribe. Formularies exist to balance cost savings and effectiveness to ensure that there is an efficiency to prescribing. It sounds harsh on the one hand, but formularies are carefully balanced to ensure that at least one drug is available for major categories of need, even if it is the generic version. Once generics for a given brand name enter the market, it is not uncommon for 80 percent of prescriptions to be transferred from the brand name to the generic form for small-molecule drugs. Why? It is often at the discretion of the insurance company as to which form will be covered per their individual formulary.[25]

To better understand the impact of formularies, let's take a look at an example. In the United States, the publicly funded Medicare Part D, the portion of Medicare pertaining to benefits for prescription drugs, has a formulary. Within the Medicare Part D formulary exist tiers that govern how much a drug will cost a patient. Tier 1 covers most generic prescription drugs and would be the low-est copayment. The copayment increases in Tier 2 (for preferred brand name prescription drugs) and further rises for Tier 3 (for nonpreferred brand name prescription drugs) before escalating further to a specialty tier for prescription drugs the formulary deems to have a prohibitively high cost.[26] Without an exist-ing formulary, Medicare Part D would be made vulnerable to prescribing that could cost it billions of dollars. Formularies for private insurance in the United States are usually reigned over by a similar tier system.

In the United Kingdom, health care is universal and publicly funded by the National Health Service (NHS). Within its borders, the National Institute for Health and Clinical Excellence, NHS Specialized Services, local guidelines, and individual hospital "trusts" jointly establish the formulary. At the individual hospital level, new medicines are added to the formulary by a regularly meeting committee of consultants, general practitioners, and pharmacists with an auto-matic acceptance of medicines given the green light at the national level.[27] To maintain fiscal security within the NHS, there exists a blacklist of medications considered off formulary. If a medication on the blacklist is prescribed, a patient must pay out of pocket to receive it unless exceptional circumstances exist.[28]

What Are "Me-Too" Drugs?

"Me-too" drugs have an apt name, as the title is often given to drugs that are synthesized, tested through clinical trials, and brought to market using an exist-ing, successful pharmaceutical as a structural model and without much concern paid to increased effectiveness compared to the original model drug. Louis S. Goodman, giant in the world of pharmacology thanks to the enduring *Goodman and Gilman's the Pharmacological Basis of Therapeutics*, first used the term in a 1956 symposium. At the symposium, Goodman described ongoing problems with pharmaceutical companies modifying their own or other companies' suc-cesses instead of paving the way forward with research and development into new therapeutics that would bring an increased benefit to the patient. Goodman specifically railed against "the problem of the introduction of 'me too' drugs, that is, drugs without signal advantage of any sort," in the symposium given to the National Academy of Sciences.[29]

Goodman's key concern about me-too drugs still echoes decades later, as the medications under this umbrella are often patented and marketed to line the pockets of pharmaceutical companies by taking advantage of previous research and development and the increased chances of approval thanks to tiny, incre-mental structural changes. This is particularly true with "me-again" drugs, which are produced and patented with slight structural changes to a popular drug going off patent.[30]

An early example of a me-too drug is the tricyclic antidepressant amitriptyline, released in 1960, which modeled itself after imipramine, released in 1958. Amitriptyline differs from imipramine only in the substitution of a carbon for a nitrogen and offered no increased benefit for the patient.[31] Interestingly enough, amitriptyline persists and is even on the World Health Organization's List of Essential Medicines for use in palliative care and depressive disorders.[32] A more modern example of a me-too or me-again drug comes in the separation of a racemic mixture of isomers, as we saw in the discussion of patent extension with the mixture of isomers in Prilosec and the separation of the mixtures to just the left-handed isomer for its successor Nexium.[33] As long as there remains low-hanging fruit and patent time lines in the world of pharmacy, me-too drugs will exist. It is our job as consumers to be informed and know that newer might not always mean better.

What Is Pharmacogenomics?

The realm of pharmacogenomics looks at an individual patient's genome or a part of the genome that is known to metabolize drugs and tailors dosage and prescriptions to the individual person based on this new information. This is particularly important in individuals who might react differently to a prescription medication, as it can prevent adverse effects.

Pharmacogenomics, at the moment, targets only a handful of genes known to code for proteins that play a role in drug metabolism. Such genes of interest include those that code for the cytochrome p450 family of enzymes, specifically the liver enzyme CYP2D6. CYP2D6 plays a role in the metabolism of nearly one-quarter of all prescription pharmaceuticals.[34] Of particular importance is the role CYP2D6 plays in the metabolism of codeine. Codeine is a pro-drug, meaning it is not active until the body modifies it through metabolic processes wherein it becomes morphine. This job of changing codeine into morphine is performed by CYP2D6. If your genetic makeup is such that you have only two copies of the gene coding for CYP2D6, you metabolize codeine in a normal manner. However, if you have more than two copies of the gene coding for CYP2D6, you will naturally have more of the CYP2D6 enzyme in your body, ready and willing to turn codeine into morphine. The more CYP2D6 enzymes available, the faster your body changes codeine into morphine, with possibly devastating effects, including overdoses from a standard dose of codeine.[35] In 2017, the FDA restricted the use of codeine and tramadol (which is converted by CYP2D6 to the opioid O-desmethyltramadol) in children due to general concerns and out of an abundance of caution regarding situations where a child may be a "ultra-rapid metabolizer" of codeine or tramadol.[36] Wide-scale sequencing of genes and the application of pharmacogenomics would prevent a variety of dangerous dosing instances from occurring by informing health care providers that a patient may have a heightened ability to produce CYP2D6 and thus should not be prescribed codeine. Pharmacogenomics would also lead to more subtle but still important changes in prescribing anything from heart medication to antidepressants.

So why are we not all rushing out to find out the secrets of our genome in hopes of ensuring better pharmaceutical health? A couple of issues resound. Privacy plays a role. Do you want your insurance company knowing your entire genetic makeup? Also, it is taking some time for health care providers to adopt the testing and implement it in their practices.[37] Wide-scale genetic testing would also be an added expense to an already encumbered health care system. As prices decrease and stigma wanes, however, pharmacogenomics will play a role in everyone's life.

What Is "Pharming," and Could It Be the Future of Pharmaceuticals?

Pharming is the process of using an already existing crop or animal to produce a pharmaceutical that is typically not a small-molecule medication. In a way, pharming is not all that different from the manufacture of biologics as we discussed earlier, but with pharming, goats and tobacco plants are used instead of *Escherichia coli* cells. These animals and plants are transgenic, as they can have DNA intentionally placed in them from an unrelated organism, often humans. Pharming can be a bit controversial, as it blurs the line between pharmaceutical and food, and this has no doubt slowed its application. Worries over cross contamination with the food supply abound, and these worries are not without merit. In an ideal situation, pharming would take place in plants that are unlikely to reach the food supply. In 2002, the company ProdiGene left seeds from their pig vaccine–producing maize in a field later used for the growth of soybeans.[38] ProdiGene was supposed to thoroughly remove all plants and seeds after use of the area. These modified maize seeds began to grow among the soybeans meant for consumption as food and not for pharmaceutical use.[39] This kerfuffle led to ProdiGene being fined $250,000 by the U.S. Department of Agriculture, with the company also tasked with the nearly $3 million effort for the destruction and burning of 155 surrounding acres of soybeans and accompanying cleanup.[40]

The most successful application of pharming thus far has come from the use of mammalian animals to produce pharmaceuticals. Why mammals?

Mammals work well because their milk is reliably produced, and the expression of the desired biologic pharmaceutical in the milk could be done with little detriment to the animal itself. The urine of the "pharm animals" could also be used, but the urea content in urine may unfold many of the desired proteins used as biological pharmaceuticals. This unfolding could threaten their efficacy and at a minimum require additional steps before the biologics could be used. To prepare a mammal for pharming, recombinant DNA is used to select for the desired biologic (akin to the cellular production of biologics). Once the recombinant DNA is inserted, the biologic of choice is expressed in the milk of the animal. The milk is then purified to recover the drug instead of the process of cell harvesting. In many cases, the recombinant DNA would also be passed on to the mammal's offspring, creating more and more pharming factories to be used in drug production. One such drug already exists thanks to the introduction of recombinant DNA into mammals: ATryn by rEVO Biologics. This drug is a form of the protein antithrombin expressed and then purified from the milk of genetically modified goats. ATryn, approved by the FDA in 2009, is used to preventively stop the formation of blood clots in patients who are genetically predisposed to clot formation and are about to undergo surgery or give birth.[41] Antithrombin is typically supplied by donations of human plasma, but a single one of rEVO's 200 goats is capable of supplying enough antithrombin for 90,000 plasma donations.[42] Running at full capacity and without the further reproduction of any goats, rEVO could supply as much antithrombin through ATryn as 18 million plasma donations, nearly one-half of the total 38 million plasma donations received in the United States in 2016.[43]

Why Do Direct-to-Consumer Pharmaceutical Advertisements Exist?

Oh hey, you're watching television. A commercial break comes along, you're checking your phone, and there is a pharmaceutical commercial for a drug with a fancy name and a mishmash of side effects plastered across serene images at the close, then another and yet another before the break is finally over and you are back to your show. At the moment, the United States and New Zealand have the laxest policies regarding direct-to-consumer pharmaceutical advertisements, as they allow the advertisements to not only name a drug and what it treats but also make claims to its effectiveness and safety.[44] New Zealand and the United States are far and away the exceptions to the rule, with most countries not allowing such invasive advertisements. Canada, for example, allows advertisements deemed "reminder ads," which mention the product name or the malady the product treats but not both in the same commercial.[45] These reminder ads act as subtle prompts for a viewer to consult their physician for additional information. The countries of the European Union took the situation a step farther by banning prescription pharmaceutical advertisements altogether in the early twenty-first century, but this ban is frequently subject to attempts to be overturned.[46] In New Zealand, health care providers have repeatedly called for a ban on direct-to-consumer advertisements but have had no luck so far.[47]

Direct-to-consumer advertising started in print form in the United States in 1981. The first advertisement was for Merck's new anti-pneumococcal vaccine Pneumovax in *Reader's Digest*. Broadcast ads came two years later in the United States, with the first touting the price benefit of Boots Pharmaceuticals' existing brand of the long-since-generic ibuprofen when compared to brand-name Motrin.[48]

A 2007 study showed the average U.S. consumer to view more than sixteen hours of prescription drug commercials a year, far more time than most spent with a health care provider in the same year.[49] This gap no doubt increased in the past decade. With so many hours of commercials, you might ask, "Why do we see so many direct-to-consumer advertisements in the United States?" The volume stems from a 1997 change in FDA rules allowing advertisements to include just the major risks associated with a pharmaceutical instead of a lengthy rundown of every individual side effect. This opened up advertising to the shorter fifteen- and thirty-second time windows of commercials and thus the deluge we often experience today.[50]

While the barrage continues, direct-to-consumer advertising is successful only if the patient is receptive. The success of direct-to-consumer pharmaceutical advertising hinges on the patient's ability to influence their health care provider's prescribing decisions by opening a dialogue wherein they ask for a particular drug by name. The outcome of the proliferation of advertisements is controversial, with arguments possible for both their positive and their negative effect on patient care. On the one hand, they prompt a person to see a physician or other health care provider. This is important, as the individual might not otherwise see a health care provider and work to alleviate a serious condition. The commercials are specifically successful in this regard when they also remove the stigma associated with a given disease, like pharmaceutical advertisements touting treatments for erectile dysfunction or genital herpes. On the other hand, one can make an argument that the commercials put undue strain on the patient–physician relationship. This occurs if and when the physician feels compelled to prescribe a drug the patient does not really need or if the patient feels they are not being heard and decides to cut off contact with the physician. [51]

Are Drug Prices Cheaper
outside of the United States?

A 2017 study of ten high-income countries (Australia, Canada, France, Germany, the Netherlands, Norway, Sweden, Switzerland, the United Kingdom, and the United States) showed the United States to pay about the same per person for medications as the other nine countries until a sharp escalation in the late 1990s.[52] Why the escalation occurred is the subject of debate, but it is interesting to note it ties in nicely with the 1997 rollback in direct-to-consumer marketing regulations. The escalation is also likely the result of the rampant growth of the pharmaceutical sector at the time and the introduction of new medications to the marketplace, as all ten countries studied show a marked increase in cost but not the meteoric rise seen in the United States over the next two decades to the present.

Examples of the differences in price between high-income countries are stark, in part due to the presence of nationally mandated insurance programs in many of the countries. With a national health care system comes the benefit of price negotiations with drug manufacturers but also the application of a nationwide formulary and cost-savings analyses. For example, a dose of the rheumatoid arthritis and Crohn's disease–fighting biological injection Humira costs on average $2,436 in the United States but only $444 in Australia and $650 in Sweden. This expanse extends to small-molecule drugs as well, with the blood thinner Eliquis priced on average at $6.98 per dose in the United States, far higher than Germany's $1.14 and the United Kingdom's price of $1.29 per dose.[53]

So why do those in the United States typically pay more for prescription medications? It is partly because the United States has a large uninsured population; in addition, those insured often pay a higher copay than their insured (whether private or nationally funded) counterparts in other high-income countries. Instead of a tiered copay system as often exists for the insured in the United States, Australia caps copays at $25 (38.30 AUD), while the United Kingdom uses a standard copay of $11.06 (£8.60) for those in England, with no copay at all for those residing in Scotland, Wales, and Northern Ireland.[54] Admittedly, these lower copays and prescription drug prices do not factor in the increased tax burden that may be on individuals living in countries with nationally funded health insurance.

Prescription drug prices are also subject to the free market in the United States rather than the checks and balances on drug prices through price negotiations that come with a nationally funded health care system. Currently, the U.S. Department of Defense (DoD) and the Veterans Administration (VA) are the only federal groups allowed to negotiate directly with pharmaceutical manufacturers, with this parleying leading to the DoD and VA paying on the order of half of what is paid at a retail pharmacy.[55] A 2019 U.S. House of Representatives report suggests that Medicare Part D could save $49 billion, nearly half of its entire 2016 expenditures, if allowed to negotiate directly in the manner of the DoD and VA.[56] Untangling prices for private insurance is a different set of equations altogether, with hope for prescription drug price decreases likely not on the horizon for this subset of the U.S. population without the implementation of a nationally funded system and its inherent bargaining power and price controls.[57]

Hopefully, this discussion leaves you feeling a little better informed about several of the mysterious quirks of the pharmaceutical landscape. If not, at least it might make you roll your eyes a little bit as you realize you are watching your fifth commercial of the night for the same drug.

Epilogue
MAKING THE COVID-19 VACCINE

This book follows a rough chronological order, save for a few chapters. Looking back on this short arc of medicinal discovery allows us to take on an interesting point of view. As humankind advanced in its abilities to develop medicines, we went from tackling the major issues of the era, such as acute pain relief, bacterial infections, and the like, to honing in on more individual and less life-disrupting problems, such as hair loss and erectile dysfunction. Are we better for it? I am not sure. This is not to say that there is no more room for new antibiotics; instead, reality is quite the opposite. We lack new, novel antibiotics that could lie in wait to protect us from unforeseen infections years down the line. Developing such a medication is often not "flashy" enough to warrant a considerable amount of attention from commercial pharmaceutical companies unless prodded by governments.

We would be remiss to cover the subject of making medicine without devoting some time to the world upheaval due to COVID-19 and the creation of a number of vaccines against it. At the time of writing, four different COVID-19 vaccines dominate most of the headlines. Three see use in the United States—Pfizer-BioNTech's messenger RNA (mRNA for short) vaccine, Moderna's mRNA vaccine, and Johnson & Johnson's virus vector vaccine. A fourth vaccine has approval in the United Kingdom and several other countries, a joint effort by Oxford University and AstraZeneca. There are a number of other vaccines available worldwide as well, including Cuba's Abdala, Russia's Sputnik V, and China's multiple offerings from Sinopharm and Sinovac.[1]

While the technology behind the vaccines may differ, at their core is the code for the COVID-19 spike protein. The spike protein makes up the little bumps you see on the now ubiquitous image of the coronavirus. The proteins make up the same physical feature that lends the "corona" part to coronavirus due to their crown-like appearance. The spike protein is also the part of the coronavirus that infiltrates the cell. By itself, the spike protein is inherently harmless, but it can be used to signal the body that a foreign material has entered into it. Unlike many of the medications covered in this book, the COVID-19 vaccines were definitely not found by accident but instead were won through enormous effort on the shoulders of an existing knowledge base.

The Pfizer-BioNTech and Moderna mRNA vaccines share a technological background. The duo makes use of the spike protein's code through a messenger RNA sequence embedded in a lipid shell. Once injected into the body, the cells use the code from the mRNA to make the spike protein. In doing so, an immune response from the body is triggered, one that hopefully persists should one come in contact with the virus in their everyday life.[2] This vaccination route requires two injections spread weeks apart to provide full vaccination two weeks after the second injection. One drawback to the Pfizer-BioNTech and Moderna offerings is that the nature of the mRNA vaccine necessitates extremely cold storage temperatures, hampering distribution in some areas.

Moderna's vaccine came into being quite quickly, with a prototype vaccine created within a week after the SARS-CoV-2 virus genome was made public. SARS-CoV-2, short for severe acute respiratory syndrome coronavirus 2, is the virus that leads to the illness COVID-19. Scientists have pined over mRNA therapies for the past four decades, with Moderna making headway in that direction since its founding in 2010.[3] Moderna's name actually has RNA in it, using the combination of "modified" and "RNA" to form the decade-old company's moniker.[4]

Unlike the Pfizer-BioNTech and Moderna mRNA vaccines, the Johnson & Johnson and Oxford/AstraZeneca vaccines place the code for the spike protein within a harmless virus. Once the virus enters the cells, the cells use the code to make the spike protein, and thus the stage is set for an immune response should there ever be an actual SARS-CoV-2 infection in the body.[5]

In the eyes of some, the Johnson & Johnson vaccine holds a major advantage, as just a single shot is necessary instead of the pair of shots required by the Pfizer-BioNTech and Moderna vaccines. The Johnson & Johnson vaccine also exhibits more relaxed storage temperature tolerances, allowing it to be used in more rural settings that may not have the cold storage facilities necessary to preserve the Pfizer-BioNTech and Moderna offerings.

There are several other COVID-19 vaccines in development or available worldwide, foremost being Russia's Sputnik V and Sputnik Light vaccines. Both are viral vector vaccines akin to the Johnson & Johnson and Oxford/AstraZeneca offerings, with Sputnik V calling for a second dose that deftly changes the viral vector of choice for the second shot. The Sputnik duo have seen distribution into more than seventy countries.[6]

How Did We Get Vaccines for COVID-19 So Quickly?

Prior to the COVID-19 vaccine, the shortest time period for a novel vaccine's development was four years. This scientific feat was achieved in the creation of a second mumps vaccine, already benefiting from a reasonably well understood disease state—mumps—with an inferior vaccine already on the market.[7] So how did we go from an explosion of worldwide COVID-19 cases to multiple working vaccines in less than a year? A couple of reasons. Operation Warp Speed, initiated by the United States, provided $10 billion to pharmaceutical companies ready to tackle the problem, along with additional billions from the World Bank and a plethora of other international organizations.[8] Such a vast pool of money allowed for pharmaceutical companies to run several parts of the design and safety process in parallel—for example, in many cases, wide-scale manufacturing of the vaccine was going on at the same time as clinical trials. The individual phases of the clinical trials were also conducted in parallel in some cases. With the widespread nature of COVID-19, these trials had a ready supply of clinical trial participants across the globe. The parallel manufacturing of the COVID-19 vaccines also aided in the rollout of the therapy, as four days after the U.S. Food and Drug Administration (FDA) approved the emergency use authorization for the Pfizer-BioNTech vaccine on December 10, 2020, health care workers in the United States were receiving the vaccine.[9] Such a speedy rollout would have been impossible without substantive funding from Operation Warp Speed and international entities. Moderna and the Oxford/AstraZeneca's contributions quickly followed. First available to health care workers and the elderly, access to the COVID vaccine in the United States and many parts of the world slowly trickled down by age-group.

Scientists stood on solid ground to fight SARS-CoV-2. Why? Many were already looking at similar coronaviruses, specifically those that lead to severe acute respiratory syndrome (SARS) and Middle East respiratory syndrome (MERS).[10] Such studies gave scientists a firm foundation to build on. When vital information in the form of the genome of the SARS-CoV-2 was released by a Chinese laboratory studying the Wuhan outbreak to the world and without restriction to the website Virological.org in January 2020, scientists were able to put their knowledge base to good use and in a timely manner.[11]

The collective COVID-19 vaccine creation effort is a remarkable feat showing exactly what humans can do when faced with a nigh insurmountable problem. But the story in the year after the vaccines were released is a little different.

The story has been more about those unwilling to get the vaccine in some parts of the world and the absolute dearth of access to the vaccine in many nations.

So exactly why are so many unwilling to receive the vaccine? A number of reasons swirled for an individual's reticence, running the gamut from religious beliefs to distrust of their government to thoughts that the vaccine was a conspiracy to connect individuals to newly rolled out 5G networks (I love a good conspiracy theory, but it is not) to simply not wanting to receive the vaccine.[12] Regardless of your willingness or unwillingness to receive the vaccine, the vaccine is likely with us for years to come. We are likely to see the coronavirus vaccine administered as something akin to an annual flu shot, if not given in combination with an annual flu shot. This would include any updates to the vaccine, much like the influenza vaccine is attenuated each year to the strains that are in circulation already along with those predicted to be most prevalent. Six World Health Organization Collaborating Centers (two in the United States, one in the United Kingdom, one in Australia, one in China, and one in Japan) choose the strains to be included in the seasonal influenza vaccine in light of data submitted from 144 national influenza centers operating in 114 countries across the globe.[13] We will likely see similar implementation if we do indeed have a recurring coronavirus vaccine in our future.

As time passes, life inches closer to normal but with some changes. As concerts and sporting events restarted in North America in the early summer of 2021, many required proof of vaccination or a recent COVID test for entrance. Masking is still ubiquitous in many public situations. Amidst the vaccine acceptance controversy that continues to rage, different variants of the COVID virus came to the forefront and threatened both the vaccinated and the unvaccinated. The highly contagious delta variant surged in many parts of the globe through the summer of 2021, putting a true sense of fear back into populations just months after many rolled back masking ordinances and social distancing guidelines. As delta faded, a new threat appeared in the omicron variant in late 2021. Late 2021 also saw the FDA authorize emergency use of the Pfizer-BioNTech vaccine for use in children ages five to eleven.

In August 2021, the Pfizer-BioNTech vaccine became the first COVID-19 vaccine to receive full approval from the FDA for use in individuals aged sixteen and older. Prior to this, the trio of vaccines available in the United States were administered under an FDA emergency use authorization. Full approval allowed for the introduction of vaccine mandates, with one of the first being a mandatory administration of the vaccine to U.S. military personnel.[14] More mandates would follow suit, with President Joseph R. Biden making use of the mandates to require all federal workers and contractors to be vaccinated, with limited exceptions, along with calling on private businesses of 100 employees or more to require vaccination or weekly testing. Rounding out this wide swath was a requirement of nearly 17 million health care workers to receive the vaccine if their hospital, clinic, or other health care establishment receives Medicare or Medicaid payments.[15] Biden's heavy-handed approach echoed the push for vaccination seen decades prior, whether it be for smallpox or polio, but it may be what it takes for the United States to get over the vaccination hump.[16] Complicating the situation further is the imminent third (or even fourth) vaccine dose for many, making for an ethical conundrum as some receive a third dose while others in the world willing to accept the vaccine have yet to receive their first.

In the year after widespread distribution of the vaccine in the United States, it has become a plodding trek toward immunity, with significant portions of the population reticent to receive the vaccine despite the dangers of COVID-19 and the very real threat from its variants. At the moment, nearly 6 billion doses of COVID-19 vaccine have been given worldwide. Malta, the United Arab Emirates, Portugal, and Singapore lead the globe in vaccinations, with more than 80 percent of their populations fully vaccinated. Canada comes in at 70 percent, with the United Kingdom at 66 percent, while the United States trails behind at a dismal 54 percent.[17]

We have come a long way from chewing bark to injecting the code for spike proteins into our bodies, and I wonder what Alexander Fleming, William Withering, Joseph Priestly, and the other luminaries we have touched on in this book would think of that.

Acknowledgments

An enormous thank-you to my wife, who has been with me through everything. Thank you for walking this road with me. Thanks to my daughters for their support in whatever it is that Daddy does. Plus the hugs, thanks for all the hugs.

My thanks go out to my mother for her sacrifices and steadfastness over the years. Thanks to my father and stepmother for their support and encouragement. Thanks to my brother as well, especially for the trips to get me out of the house. Thanks to my mother-in-law for her support throughout the years. Thanks to my family and friends for their many efforts in my life and ongoing encouragement. I would also like to thank Laurie Norton, Rhea Wynn, Lucinda Savage, Carol Pigg, Brenda McCroskey, Susan Picken, Kent Clinger, James Ronald Boone, Aaron Lucius, Wayne Garrett, William Tallon, and the other educators who guided me throughout my years. Thanks also to all of the people I came in contact with at io9.

I wish to extend my appreciation to Laura Wood at FinePrint Literary Management for her perseverance and the effort she expended, as well as Jake Bonar, Nicole Carty, Janine Faust, Karen Weldon, and Bruce Owens at Rowman & Littlefield for their assistance in the writing process and editing. Thanks also to the good people at Globe Pequot and Prometheus for their work in the design and distribution of this book. Finally, I thank you, especially if you've read this far.

Notes

Introduction

1. Sam Michael et al., "A Robotic Platform for Quantitative High-Throughput Screening," *Assay and Drug Development Technologies* 6, no. 5 (2008): 637–57, doi: 10.1089/adt.2008.150.

2. Philine Kirsch et al., "Concepts and Core Principles of Fragment-Based Drug Design," *Molecules* 24, no. 23 (November 2019): 1–22, doi:10.3390/molecules24234309.

3. Kirsch et al., "Concepts and Core Principles of Fragment-Based Drug Design."

4. Joe Palca, "Snail Venom Yields Potent Painkiller, but Delivering the Drug Is Tricky," National Public Radio, August 3, 2015, https://www.npr.org/sections/health -shots/2015/08/03/428990755/snail-venom-yields-potent-painkiller-but-delivering-the -drug-is-tricky.

5. Laurence L. Brunton et al., *Goodman and Gilman's The Pharmacological Basis of Therapeutics*, 11th ed. (New York: McGraw-Hill, 2006), 4.

6. C. A. Lipinski et al., "Experimental and Computational Approaches to Estimate Solubility and Permeability in Drug Discovery and Development Settings," *Advanced Drug Delivery Reviews* 46 (March 2001): 3–26.

7. "Live Attenuated Influenza Vaccine [LAIV] (The Nasal Spray Flu Vaccine)," Centers for Disease Control and Prevention, last modified September 2, 2020, https://www .cdc.gov/flu/prevent/nasalspray.htm.

8. Izet Masic et al., "Contribution of Arabic Medicine and Pharmacy to the Development of Health Care Protection in Bosnia and Herzegovina—The First Part," *Medical Archives (Sarajevo, Bosnia and Herzegovina)* 71, no. 5 (2017): 366, doi:10.5455/medarh .2017.71.364-372.

9. "Trichloroethylene, Tetrachloroethylene, and Some Other Chlorinated Agents: Chloral and Chloral Hydrate," *Monographs on the Evaluation of Carcinogenic Risks to Humans*, no. 106, accessed October 12, 2020, https://www.ncbi.nlm.nih.gov/books /NBK294283.

10. Thomas Dormandy, *The Worst of Evils* (New Haven, CT: Yale University Press, 2006), 348.

11. Howard Markel, "Marilyn Monroe and the Prescription Drugs That Killed Her," *PBS News Hour*, August 5, 2016, https://www.pbs.org/newshour/health/marilyn-monroe-and-the-prescription-drugs-that-killed-her.

12. James A. Inciardi, "The Changing Life of Mickey Finn: Some Notes on Chloral Hydrate Down through the Ages," *Journal of Popular Culture* 11, no. 3 (Winter 1977): 591.

13. Clay Cansler, "Distillations: Where's the Beef?," Science History Institute, December 13, 2013, https://www.sciencehistory.org/distillations/article/where%E2%80%99s-beef.

14. Brian J. Ford, "Critical Focus: Solving the Mystery of Spontaneous Human Combustion," *Microscope* 60, no. 2 (2012): 65.

15. "Milestones in U.S. Food and Drug Law History," U.S. Food and Drug Administration, accessed September 29, 2020, https://www.fda.gov/about-fda/fdas-evolving-regulatory-powers/milestones-us-food-and-drug-law-history.

16. "Part I: The 1906 Food and Drugs Act and Its Enforcement," U.S. Food and Drug Administration, accessed September 29, 2020, https://www.fda.gov/about-fda/fdas-evolving-regulatory-powers/part-i-1906-food-and-drugs-act-and-its-enforcement.

17. "History of FDA's Internal Organization," U.S. Food and Drug Administration, accessed September 29, 2020, https://www.fda.gov/about-fda/history-fdas-fight-consumer-protection-and-public-health/history-fdas-internal-organization.

18. "Milestones in U.S. Food and Drug Law History."

19. "Part II: 1938, Food, Drug, Cosmetic Act," U.S. Food and Drug Administration, accessed September 29, 2020, https://www.fda.gov/about-fda/fdas-evolving-regulatory-powers/part-ii-1938-food-drug-cosmetic-act.

20. "Milestones in U.S. Food and Drug Law History."

21. Ingrid Peritz, "Canadian Doctor Averted Disaster by Keeping Thalidomide Out of the U.S.," *Globe and Mail*, November 24, 2014, https://www.theglobeandmail.com/news/national/canadian-doctor-averted-disaster-by-keeping-thalidomide-out-of-the-us/article21721337.

22. "Step 3: Clinical Research," U.S. Food and Drug Administration, last modified January 4, 2018, https://www.fda.gov/patients/drug-development-process/step-3-clinical-research.

23. "Step 3."

24. "Step 3."

25. "Step 3."

26. "Step 4: FDA Drug Review," U.S. Food and Drug Administration, last modified September 29, 2020, https://www.fda.gov/patients/drug-development-process/step-4-fda-drug-review.

27. "Development and Approval Process (CBER)," U.S. Food and Drug Administration, last modified June 25, 2020, https://www.fda.gov/vaccines-blood-biologics/development-approval-process-cber.

28. "Step 3."

29. Julie Frearson and Paul Wyatt, "Drug Discovery in Academia: The Third Way?" *Expert Opinion on Drug Discovery* 5, no. 10 (2010): 909–19, doi:10.1517/17460441.2 010.506508.

30. Frearson and Wyatt, "Drug Discovery in Academia."

31. Keith D. Tait, "Chapter 79: Pharmaceutical Industry," in *Encyclopedia of Occupational Health and Safety*, 4th ed. (Geneva: International Labour Office), accessed October 6, 2020, http://www.ilocis.org/documents/chpt79e.htm.

32. Dennis D. Zaebst, "Effects of Synthetic Oestrogens on Pharmaceutical Workers: A United States Example. Chapter 79: Pharmaceutical Industry," in *Encyclopedia of Occupational Health and Safety*, accessed October 6, 2020, http://www.ilocis.org/docu ments/chpt79e.htm.

33. J. Berger et al., "How Drug Life-Cycle Management Patent Strategies May Impact Formulary Management," *American Journal of Managed Care* 22, no. 16, supplement (October 2016): S487–95, PMID: 28719222.

34. "Measuring the Risks and Rewards of Drug Development: New Research from MIT Shows That the Success Rates of Clinical Trials Are Higher Than Previously Thought," MIT Sloan Office of Media Relations, January 31, 2018, https://mitsloan .mit.edu/press/measuring-risks-and-rewards-drug-development-new-research-mit -shows-success-rates-clinical-trials-are-higher-previously-thought.

35. Matthew Herper, "The Truly Staggering Cost of Inventing New Drugs," *Forbes Healthcare*, February 10, 2012, https://www.forbes.com/sites/matthewherper/2012 /02/10/the-truly-staggering-cost-of-inventing-new-drugs/#13d953ab4a94.

36. Berger et al., "How Drug Life-Cycle Management Patent Strategies May Impact Formulary Management."

37. Chittaranjan Andrade, "Desvenlafaxine," *Indian Journal of Psychiatry* 51, no. 4 (2009): 320–23, doi:10.4103/0019-5545.58303.

38. Ullaa Lepola et al., "Do Equivalent Doses of Escitalopram and Citalopram Have Similar Efficacy? A Pooled Analysis of Two Positive Placebo-Controlled Studies in Major Depressive Disorder," *International Clinical Psychopharmacology* 19, no. 3 (May 2004): 149–55.

39. W. J. Burke et al., "Fixed-Dose Trial of the Single Isomer SSRI Escitalopram in Depressed Outpatients," *Journal of Clinical Psychiatry* 63, no. 4 (April 2002): 331–36, doi:10.4088/jcp.v63n0410, PMID: 12000207.

40. Waheed Asghar et al., "Comparative Efficacy of Esomeprazole and Omeprazole: Racemate to Single Enantiomer Switch," *Daru: Journal of Faculty of Pharmacy, Tehran University of Medical Sciences* 23, no. 50 (November 2015), doi:10.1186/s40199-015 -0133-6.

41. Joshua P. Cohen et al., "Switching Prescription Drugs to Over the Counter," *British Medical Journal (Clinical Research Edition)* 330, no. 7481 (2005): 39–41, doi:10 .1136/bmj.330.7481.39.

42. Alessandra Cristina Santos Akkari et al., "Pharmaceutical Innovation: Differences between Europe, USA, and 'Pharmerging' Countries," *Gestão & Produção* 23, no. 2 (June 14, 2016): 369–73, http://dx.doi.org/10.1590/0104-530x2150-15.

43. Wendy H. Schacht and John R. Thomas, "The Hatch–Waxman Act: Legislative Changes in the 108th Congress Affecting Pharmaceutical Patents," *Congressional Research Service Reports*, April 30, 2004, 1.

44. Martha M. Rumore, "The Hatch-Waxman Act 25 Years Later: Keeping the Pharmaceutical Scales Balanced," *Pharmacy Times*, August 2009, https://www.pharmacy times.com/publications/supplement/2009/GenericSupplement0809/Generic-Hatch Waxman-0809.

45. Ryan Conrad and Randall Lutter, "New Evidence Linking Greater Generic Competition and Lower Generic Drug Prices," *U.S. Food and Drug Administration Report*, December 2019, 2–3, https://www.fda.gov/media/133509/download.

46. William C. Lamanna et al., "Maintaining Consistent Quality and Clinical Performance of Biopharmaceuticals," *Expert Opinion on Biological Therapy* 18, no. 4 (2018): 369–79, doi:10.1080/14712598.2018.1421169.

47. Schacht and Thomas, "The Hatch Waxman Act," 2.

48. Rumore, "The Hatch-Waxman Act 25 Years Later."

49. Rumore, "The Hatch-Waxman Act 25 Years Later."

50. Brendan Pierson, "U.S. Court Upholds AstraZeneca, Ranbaxy Win in Nexium Antitrust Trial," Reuters, November 21, 2016, https://www.reuters.com/article/astra zeneca-nexium-appeal/u-s-court-upholds-astrazeneca-ranbaxy-win-in-nexium-antitrust -trial-idUSL1N1DM1W2.

Chapter 1

1. "Charming but Fanciful: The Fleming-Churchill Myth," Churchill Project, Hillsdale College, last modified September 25, 2018, https://winstonchurchill.hillsdale.edu /alexander-fleming-saved-churchill.

2. Siang Yong Tan and Yvonne Tatsumura, "Alexander Fleming (1881–1955): Discoverer of Penicillin," *Singapore Medical Journal* 56, no. 7 (July 2015): 366, doi:10.11622/smedj.2015105.

3. Tan and Tatsumura, "Alexander Fleming (1881–1955)," 366.

4. "Sir Alexander Fleming: Biographical," *Nobel Media AB* (2020), accessed October 12, 2020, https://www.nobelprize.org/prizes/medicine/1945/fleming/biographical.

5. David John Davis, "Bacteriology and the War," *Scientific Monthly* 5, no. 5 (November 1917): 393–94, https://www.jstor.org/stable/22551.

6. Alexander Fleming, "The Physiological and Antiseptic Action of Flavine (with Some Observations on the Testing of Antiseptics)," *Lancet* 190, no. 4905 (September 1, 1917): 341–45, https://doi.org/10.1016/S0140-6736(01)52126-1.

7. Alexander Fleming, "On a Remarkable Bacteriolytic Element Found in Tissues and Secretions," *Proceedings of the Royal Society of London B* 93 (1922): 306–17, https://doi .org/10.1098/rspb.1922.0023.

8. Jolanta Lukasiewicz and Czeslaw Lugowski, "Editorial: O-Specific Polysaccharide Confers Lysozyme Resistance to Extraintestinal Pathogenic Escherichia Coli." *Virulence* 9, no. 1 (2018): 919, doi:10.1080/21505594.2018.1460188.

9. "Look East: Alexander Fleming's Country Home," *East Anglian Film Archive*, accessed October 13, 2020, http://www.eafa.org.uk/catalogue/5535.

10. Alexander Fleming, "On the Antibacterial Action of Cultures of a Penicillium, with Special Reference to Their Use in the Isolation of B. Influenzæ," *British Journal of Experimental Pathology* 10, no. 3 (May 1929): 226.

11. Fleming, "On the Antibacterial Action of Cultures of a Penicillium."

12. Alexander Fleming, "Penicillin," Nobel lecture, December 11, 1945, 83, https://www.nobelprize.org/prizes/medicine/1945/fleming/lecture.

13. Fleming, "On the Antibacterial Action of Cultures of a Penicillium."

14. Ronald Hare, "New Light on the History of Penicillin," *Medical History* 26, no. 1 (1982): 5, doi:10.1017/s0025727300040758.

15. Fleming, "Penicillin," 89.

16. "First Use of Penicillin at University of Sheffield Recognized in the UK's Best Breakthroughs List," University of Sheffield, December 6, 2018, https://www.sheffield.ac.uk/news/nr/first-use-pencillin-at-sheffield-best-breakthrough-1.820181.

17. Ronnie Henry, "Etymologia: Penicillin," *Emerging Infectious Diseases* 25, no. 1 (January 2019): 62, doi:10.3201/eid2501.ET2501.

18. Milton Wainwright and Harold T. Swan, "C. G. Paine and the Earliest Surviving Clinical Records of Penicillin Therapy," *Medical History* 30 (1986): 42–56.

19. David S. Jones and John H. Jones, "Sir Edward Penley Abraham CBE," *Biographical Memoirs of Fellows of the Royal Society*, no. 60 (2014): 8.

20. Jones and Jones, "Sir Edward Penley Abraham CBE," 9–11.

21. "Howard Walter Florey and Ernst Boris Chain," Science History Institute, last modified December 4, 2017, https://www.sciencehistory.org/historical-profile/howard-walter-florey-and-ernst-boris-chain.

22. Marc A. Shampo and Robert A. Kyle, "Ernst Chain—Nobel Prize for Work on Penicillin," *Mayo Clinic Proceedings* 75, no. 9 (September 2000): 882.

23. C. L. Moberg, "Penicillin's Forgotten Man: Norman Heatley," *Science* 253, no. 5021 (August 1991): 734.

24. Eric Sidebottom, "Obituaries: Norman Heatley," *Independent*, January 23, 2004, https://www.independent.co.uk/news/obituaries/norman-heatley-37866.html.

25. Ernest Chain et al., "Penicillin as a Chemotherapeutic Agent," *Lancet* 236, no. 6104 (August 24, 1940): 226–28.

26. Howard Markel, "The Real Story behind Penicillin," *PBS News Hour*, September 27, 2013, https://www.pbs.org/newshour/health/the-real-story-behind-the-worlds-first-antibiotic.

27. Tom Calver, "75 Years of Penicillin in People," University of Oxford, Oxford Science Blog, February 12, 2016, https://www.ox.ac.uk/news/science-blog/75-years-penicillin-people.

28. Markel, "The Real Story behind Penicillin."

29. Robert Gaynes, "The Discovery of Penicillin—New Insights after More Than 75 Years of Clinical Use," *Emerging Infectious Diseases* 23, no. 5 (May 2017): 850.

30. Wolfgang Saxon, "Anne Miller, 90, First Patient Who Was Saved by Penicillin," *New York Times*, June 9, 1999, https://www.nytimes.com/1999/06/09/us/anne-miller-90-first-patient-who-was-saved-by-penicillin.html.

31. Lily Rothman, "This Is What Happened to the First American Treated with Penicillin," *Time*, March 14, 2016, https://time.com/4250235/penicillin-1942-history.

32. J. H. Humphrey, "Excretion of Penicillin in Man," *Nature* 154, no. 765 (1944): 765, https://doi.org/10.1038/154765a0.

33. Rebecca Kreston, "The Magic Arrow: Penicillin and the Recurrin' Urine," *Discover,* December 31, 2014, https://www.discovermagazine.com/health/the-magic-arrow -penicillin-and-the-recurrin-urine.

34. Peter Andrey Smith, "How the Largest Nightclub Fire in US History Became a Milestone in Modern Medicine," *Business Insider*, October 29, 2017, https://www .businessinsider.com/nightclub-fire-became-a-milestone-in-modern-medicine-2017-10.

35. Ernst B. Chain, "The Chemical Structure of the Penicillins," Nobel lecture, March 20, 1946, 1–2, https://www.nobelprize.org/prizes/medicine/1945/chain/lecture.

36. "Penicillin: Opening the Era of Antibiotics," National Center for Agricultural Utilization Research, U.S. Department of Agriculture, last modified May 3, 2018, https:// www.ars.usda.gov/midwest-area/peoria-il/national-center-for-agricultural-utilization -research/docs/penicillin-opening-the-era-of-antibiotics.

37. Phil Luciano, "Peoria Played Key Role in Mass Production of Penicillin," *Peoria Journal Star*, April 25, 2019, https://www.pjstar.com/special/20190425/peoria-played -key-role-in-mass-production-of-penicillin.

38. Howard Markel, "The Real Story behind Penicillin."

39. Derek W. Yip and Valerie Gerriets, "Penicillin," *StatPearls*, last modified May 12, 2020, https://www.ncbi.nlm.nih.gov/books/NBK554560.

40. Yip and Gerriets, "Penicillin."

41. Edward P. Abraham and Ernst Chain, "An Enzyme from Bacteria Able to Destroy Penicillin," *Nature* 146, no. 3713 (December 1940): 837.

42. "American Chemical Society International Historic Chemical Landmarks. Discovery and Development of Penicillin," American Chemical Society, accessed October 15, 2020, https://www.acs.org/content/acs/en/education/whatischemistry/landmarks /flemingpenicillin.html.

43. Roswell Quinn, "Rethinking Antibiotic Research and Development: World War II and the Penicillin Collaborative," *American Journal of Public Health* 103, no. 3 (March 2013): 426–34, doi:10.2105/AJPH.2012.300693.

44. "American Chemical Society International Historic Chemical Landmarks. Discovery and Development of Penicillin," The American Chemical Society, accessed October 15, 2020, https://www.acs.org/content/acs/en/education/whatischemistry/land marks/flemingpenicillin.html.

45. Tom Calver, "75 Years of Penicillin in People."

46. Gaynes, "The Discovery of Penicillin," 849–53.

47. Joachim Pietzsch, "Enhancing X-Ray Vision," Nobel Prize, accessed October 15, 2020, https://www.nobelprize.org/prizes/chemistry/1964/perspectives.

48. David Greenwood, *Antimicrobial Drugs: Chronicle of a Twentieth Century Medical Triumph* (Oxford: Oxford University Press, 2008), 121.

49. John C. Sheehan and Kenneth R. Henery-Logan, "The Total Synthesis of Penicillin V," *Journal of the American Chemical Society* 79, no. 5 (1957): 1262–63.

50. Rothman, "This Is What Happened to the First American Treated with Penicillin."

51. Moberg, "Penicillin's Forgotten Man," 734.

52. Eric Sidebottom, "Obituaries."

53. Moberg, "Penicillin's Forgotten Man," 734.

54. Fleming, "Penicillin."

55. James Badcock, "Anti-Bullfight Activists Accidentally Vandalise Madrid Statue of Penicillin Inventor Alexander Fleming," *Telegraph*, July 1, 2020, https://www.telegraph .co.uk/news/2020/07/01/anti-bullfight-activists-accidentally-vandalise-madrid-statue.

Chapter 2

1. J. Achan et al., "Quinine, an Old Anti-Malarial Drug in a Modern World: Role in the Treatment of Malaria," *Malaria Journal* 10, no. 144 (2011): 1, https://doi.org /10.1186/1475-2875-10-144.

2. Alec Haggis, "Fundamental Errors in the Early History of Cinchona: The Fabulous Story of the Countess of Chinchon," *Bulletin of the History of Medicine* 10, no. 3 and 4 (October and November 1941): 568–87.

3. Fiammetta Rocco, *The Miraculous Fever Tree: Malaria and the Quest for a Cure That Changed the World* (New York: HarperCollins, 2003), 70, 78.

4. L. J. Bruce-Chwatt, "Three Hundred and Fifty Years of the Peruvian Fever Bark," *British Medical Journal (Clinical Research Edition)* 296 (May 1988): 1486, https://doi .org/10.1136/bmj.296.6635.1486.

5. G. Gachelin et al., "Evaluating Cinchona Bark and Quinine for Treating and Preventing Malaria," *Journal of the Royal Society of Medicine* 110, no. 1 (January 2017): 32, doi:10.1177/0141076816681421.

6. T. W. Keeble, "A Cure for the Ague: The Contribution of Robert Talbor (1642–81)," *Journal of the Royal Society of Medicine* 90, no. 5 (May 1997): 287–88, doi:10.117 7/014107689709000517.

7. Paul Reiter, "From Shakespeare to Defoe: Malaria in England in the Little Ice Age," *Emerging Infectious Diseases* 6, no. 1 (January–February 2000): 6, doi:10.3201 /eid0601.000101.

8. John N. Wood, "From Plant Extract to Molecular Panacea: A Commentary on Stone (1763) 'An Account of the Success of the Bark of the Willow in the Cure of the Agues,'" *Philosophical Transactions of the Royal Society of London, Series B, Biological Sciences* 370, no. 1666 (April 2015): 2, doi:10.1098/rstb.2014.0317.

9. Bruce-Chwatt, "Three Hundred and Fifty Years of the Peruvian Fever Bark."

10. Wood, "From Plant Extract to Molecular Panacea."

11. "Why Do Mosquitoes Bite Me and Not My Friend?" Library of Congress, accessed October 19, 2020, https://www.loc.gov/everyday-mysteries/item/why-do-mosquitoes -bite-me-and-not-my-friend.

12. G. Dennis Shanks, "Historical Review: Problematic Malaria Prophylaxis with Quinine," *American Journal of Tropical Medicine and Hygiene* 95, no. 2 (August 3, 2016): 269, doi:10.4269/ajtmh.16-0138.

13. "About Malaria," Centers for Disease Control and Prevention, last modified September 17, 2020, https://www.cdc.gov/malaria/about/faqs.html.

14. "About Malaria: Uncomplicated Malaria," Centers for Disease Control and Prevention, last modified January 9, 2019, https://www.cdc.gov/malaria/about/disease.html.

15. "About Malaria."

16. "Malaria," World Health Organization, accessed October 19, 2020, https://www.who.int/news-room/fact-sheets/detail/malaria.

17. K. J. Arrow, ed., *Saving Lives, Buying Time: Economics of Malaria Drugs in an Age of Resistance* (Washington: National Academies Press, 2004), 125.

18. Robert Sallares et al., "The Spread of Malaria to Southern Europe in Antiquity: New Approaches to Old Problems," *Medical History* 48, no. 3 (July 2004): 314, doi:10.1017/s0025727300007651.

19. Andrew Thompson, "Malaria and the Fall of Rome," BBC, February 2, 2011, https://www.bbc.co.uk/history/ancient/romans/malaria_01.shtml.

20. Charles C. Mann, *1493: Uncovering the New World Columbus Created* (New York: Knopf, 2011), 90.

21. Mann, *1493*, 116.

22. Mann, *1493*, 116.

23. J. V. Pai-Dhungat, "Caventou, Pelletier &—History of Quinine," *Journal of the Association of Physicians of India* 63, no. 3 (March 2015): 68.

24. L. F. Hass, "Pierre Joseph Pelletier (1788–1842) and Jean Bienaime Caventou (1795–1887)," *Journal of Neurology, Neurosurgery and Psychiatry* 57 (1994): 1333, doi:10.1136/jnnp.57.11.1333.

25. Achan et al., "Quinine, an Old Anti-Malarial Drug in a Modern World," 1.

26. Sergey Kapishnikov et al., "Unraveling Heme Detoxification in the Malaria Parasite by In Situ Correlative X-Ray Fluorescence Microscopy and Soft X-Ray Tomography," *Nature Scientific Reports* 7, no. 7610 (2017): 1–2, https://doi.org/10.1038/s41598-017-06650-w.

27. Elizabeth Davis, "Dr. Sappington's Anti-Fever Pills," *Columbia Daily Tribune*, June 1, 2020, https://www.columbiatribune.com/story/news/columns/2020/06/01/dr-sappingtons-anti-fever-pills/42145205.

28. Lynn Morrow, "Dr. John Sappington, Southern Patriarch in the New West," *Missouri Historical Review* 90, no. 1 (October 1995): 48–50, https://digital.shsmo.org/digital/collection/mhr/id/47714.

29. Achan et al., "Quinine, an Old Anti-Malarial Drug in a Modern World," 1–2.

30. Robert D. Hicks, "Distillations: 'The Popular Dose with Doctors': Quinine and the American Civil War," Science History Institute, December 6, 2013, https://www.sciencehistory.org/distillations/the-popular-dose-with-doctors-quinine-and-the-american-civil-war.

31. Shanks, "Historical Review," 269.

32. "Forces for Change: Discovery of Quinine," BBC World Service, accessed October 20, 2020, https://www.bbc.co.uk/worldservice/africa/features/storyofafrica/11generic1 .shtml.

33. Tom Cassauwers, "The Global History of Quinine, the World's First Anti-Malaria Drug," *Medium*, December 30, 2015, https://medium.com/@tcassauwers/the -global-history-of-the-world-s-first-anti-malaria-drug-d1e11f0ba729.

34. Arjo Roersch Van Der Hoogte and Toine Pieters, "Quinine, Malaria, and the Cinchona Bureau: Marketing Practices and Knowledge Circulation in a Dutch Trans-oceanic Cinchona–Quinine Enterprise (1920s–30s)," *Journal of the History of Medicine and Allied Sciences* 71, no. 2 (April 2016): 197–225, https://www.ncbi.nlm.nih.gov /pmc/articles/PMC4887601.

35. Achan et al., "Quinine, an Old Anti-Malarial Drug in a Modern World," 7.

36. Vassiliki Betty Smocovitis, "The Timeline: Desperately Seeking Quinine," *Modern Drug Discovery* 6, no. 5 (May 2003): 57.

37. Shanks, "Historical Review," 271.

38. Smocovitis, "The Timeline," 58.

39. "Cinchona Missions Expedition," Smithsonian National Museum of Natural History, accessed October 15, 2020, https://naturalhistory.si.edu/research/botany /about/historical-expeditions/cinchona-missions.

40. Nicolás Cuvi, "The Cinchona Program (1940–1945): Science and Imperialism in the Exploitation of a Medicinal Plant," *Dynamis* 31, no. 1 (2011): 197–98, doi:10.4321 /s0211-95362011000100009.

41. "History of Lupus," Columbia Doctors, Columbia University Irving Medical Center, accessed October 16, 2020, https://www.columbiadoctors.org/specialties/rheuma tology/overview/lupus-center/history-lupus.

42. Ewa Haładyj et al., "Antimalarials—Are They Effective and Safe in Rheumatic Diseases?" *Reumatologia* 56, no. 3 (2018): 167, doi:10.5114/reum.2018.76904.

43. "Parasites-Babesiosis," Centers for Disease Control and Prevention, last modified October 30, 2019, https://www.cdc.gov/parasites/babesiosis/health_professionals/index .html.

44. "CFR, Code of Federal Regulations Title 21," U.S. Food and Drug Administration, last modified April 1, 2019, https://www.accessdata.fda.gov/scripts/cdrh/cfdocs /cfcfr/CFRSearch.cfm?fr=310.546.

45. Janelle Derbis, "Serious Risks Associated with Using Quinine to Prevent or Treat Nocturnal Leg Cramps," *FDA News for Health Professionals*, September 2012, https:// www.fda.gov/media/84506/download.

46. "Food Additive Status List," U.S. Food and Drug Administration, last modified October 24, 2019, https://www.fda.gov/food/food-additives-petitions/food-additive -status-list.

47. Vittoria Traverso, "The Tree That Changed the World Map," BBC, May 28, 2020, http://www.bbc.com/travel/story/20200527-the-tree-that-changed-the-world-map.

48. P. Bacon, "Blindness from Quinine Toxicity," *British Journal of Ophthalmology* 72, no. 3 (March 1988): 219, doi:10.1136/bjo.72.3.219.

49. E. T. Sheehan et al., "Quinine and the ABCs of Long QT: A Patient's Misfortune with Arthritis, (Alcoholic) Beverages, and Cramps," *Journal of General Internal Medicine* 31 (2016): 1254–57, https://doi.org/10.1007/s11606-016-3738-7.

50. Shanks, "Historical Review," 270.

51. R. B. Woodward and W. E. Doering, "The Total Synthesis of Quinine," *Journal of the American Chemical Society* 66, no. 5 (1944): 849.

52. Katharine Sanderson, "1940s Re-Enactment Helps to Verify Old Chemical Claim," *Nature*, February 4, 2008, doi:10.1038/news.2008.554.

53. Achan et al., "Quinine, an Old Anti-Malarial Drug in a Modern World," 2–3.

54. M. D. Chowdhury et al., "A Rapid Systematic Review of Clinical Trials Utilizing Chloroquine and Hydroxychloroquine as a Treatment for COVID-19," *Academic Emergency Medicine* 27, no. 6 (2020): 493–504, doi:10.1111/acem.14005.

55. Neil Vigdor, "Man Fatally Poisons Himself while Self-Medicating for Coronavirus, Doctor Says," *New York Times*, last modified April 24, 2004, https://www.nytimes.com/2020/03/24/us/chloroquine-poisoning-coronavirus.html.

56. "FDA Cautions against Use of Hydroxychloroquine or Chloroquine for COVID-19 outside of the Hospital Setting or a Clinical Trial Due to Risk of Heart Rhythm Problems," U.S. Food and Drug Administration, last modified July 1, 2020, https://www.fda.gov/drugs/drug-safety-and-availability/fda-cautions-against-use-hydroxychloroquine-or-chloroquine-covid-19-outside-hospital-setting-or.

Chapter 3

1. G. Tsoucalas et al., "Travelling through Time with Aspirin, a Healing Companion," *European Journal of Inflammation* 9, no. 1 (January 2011): 16, https://doi.org/10.1177/1721727X1100900102.

2. Alan Cassels, "Most of Our Prescription Drugs Are Manufactured Overseas—But Are They Safe?" *Canadian Medical Association Journal* 184, no. 14 (October 2, 2012): 1648, doi:10.1503/cmaj.120416.

3. J. G. Mahdi et al., "The Historical Analysis of Aspirin Discovery, Its Relation to the Willow Tree and Antiproliferative and Anticancer Potential," *Cell Proliferation* 39, no. 2 (April 2006): 148, doi:10.1111/j.1365-2184.2006.00377.x.

4. Tsoucalas et al., "Travelling through Time with Aspirin," 13.

5. Blake de Pastino, "1,300-Year-Old Pottery Found in Colorado Contains Ancient 'Natural Aspirin,'" *Western Digs*, December 31, 2015, http://westerndigs.org/prehistoric-pottery-found-in-colorado-contains-ancient-natural-aspirin.

6. Michael J. R. Desborough, "The Aspirin Story—From Willow to Wonder Drug," *British Journal of Haematology* 177, no. 5 (2017): 674, doi:10.1111/bjh.14520.

7. Jacob Roberts, "Distillations: Sickening Sweet," Science History Institute, December 8, 2015, https://www.sciencehistory.org/distillations/magazine/sickening-sweet.

8. P. J. Didapper, "Beavers and Drug Discovery," *Pharmaceutical Journal*, May 30, 2012, https://www.pharmaceutical-journal.com/news-and-analysis/opinion/blogs/beavers-and-drug-discovery/11102128.blog?firstPass=false.

9. John N. Wood, "From Plant Extract to Molecular Panacea: A Commentary on Stone (1763) 'An Account of the Success of the Bark of the Willow in the Cure of the Agues,'" *Philosophical Transactions of the Royal Society of London, Series B, Biological Sciences* 370, no. 1666 (April 2015): 3, doi:10.1098/rstb.2014.0317.

10. Wood, "From Plant Extract to Molecular Panacea," 2–3.

11. Mahdi et al., "The Historical Analysis of Aspirin Discovery, Its Relation to the Willow Tree and Antiproliferative and Anticancer Potential," 149.

12. Desborough, "The Aspirin Story," 674.

13. Desborough, "The Aspirin Story," 674.

14. Wood, "From Plant Extract to Molecular Panacea," 3.

15. Desborough, "The Aspirin Story," 675.

16. "History: The Early Years (1863–1881)," Bayer Global, accessed October 23, 2020, https://www.bayer.com/en/history/1863-1881.

17. T. J. Rinsema, "One Hundred Years of Aspirin," *Medical History* 43, no. 4 (October 1999): 503, doi:10.1017/s0025727300065728.

18. Rinsema, "One Hundred Years of Aspirin," 504.

19. Kat Eschner, "Aspirin's Four-Thousand-Year History," *Smithsonian Magazine*, August 10, 2017, https://www.smithsonianmag.com/smart-news/brief-history-aspirin -180964329.

20. Walter Sneader, "The Discovery of Aspirin: A Reappraisal," *British Medical Journal* 321, no. 7276 (December 23, 2000): 1592, doi:10.1136/bmj.321.7276.1591.

21. Desborough, "The Aspirin Story," 6765.

22. Sneader, "The Discovery of Aspirin," 1593.

23. Kenneth C. Cummings, "Aspirin: 4,000 Years and Still Learning," *Cleveland Clinic Journal of Medicine* 86, no. 8 (August 2019): 522, doi:10.3949/ccjm.86a.19041.

24. Diarmud Jeffreys, *Aspirin: The Remarkable Story of a Wonder Drug* (New York: Bloomsbury, 2005), 73.

25. Jeffreys, *Aspirin*, 74.

26. Sneader, "The Discovery of Aspirin," 1593.

27. Desborough, "The Aspirin Story," 677.

28. Sneader, "The Discovery of Aspirin," 1593.

29. Rinsema, "One Hundred Years of Aspirin," 506.

30. Sneader, "The Discovery of Aspirin," 1593.

31. John Olmsted III, "Synthesis of Aspirin: A General Chemistry Experiment," *Journal of Chemical Education* 75, no. 10 (October 1998): 1261–63.

32. David Randall, "Focus: Aspirin—The Secret History of a Wonder Drug," *Independent*, April 17, 2005, https://www.independent.co.uk/life-style/health-and-families /health-news/focus-aspirin-the-secret-history-of-a-wonder-drug-489558.html.

33. Randall, "Focus."

34. Daniel R. Goldberg, "Aspirin: Turn-of-the-Century Miracle Drug," Science History Institute, June 2, 2009, https://www.sciencehistory.org/distillations/aspirin-turn-of -the-century-miracle-drug.

35. Charles C. Mann, *The Aspirin Wars: Money, Medicine, and 100 Years of Rampant Competition* (New York: Knopf, 1991): 40.

36. Jeffreys, *Aspirin*, 112.

37. Mann, *The Aspirin Wars*, 38–41.

38. Jeffreys, *Aspirin*, 113.

39. Esther Inglis-Arkell, "The Great Phenol Plot Is Another Reason to Hate Thomas Edison," *io9*, September 10, 2013, https://io9.gizmodo.com/the-great-phenol-plot-is-another-reason-to-hate-thomas-1279864505.

40. "Spanish Influenza Aspirin Scare," History Engine, accessed October 22, 2020, https://historyengine.richmond.edu/episodes/view/5222.

41. "Spanish Influenza Aspirin Scare."

42. Joseph C. Collins and John R. Gwilt, "The Life Cycle of Sterling Drug, Inc.," *Bulletin for the History of Chemistry* 25, no. 1 (2000): 23.

43. Tsunetoshi Shimazu, "Rapid Response: Aspirin May Be the Enhancer of Virulence in 1918 Pandemic." *British Medical Journal* 338 (May 2009): 1962, https://doi.org/10.1136/bmj.b1962.

44. Karen M. Starko, "Salicylates and Pandemic Influenza Mortality, 1918–1919 Pharmacology, Pathology, and Historic Evidence," *Clinical Infectious Diseases* 49, no. 9 (November 15, 2009): 1405–10, https://doi.org/10.1086/606060.

45. Collins and Gwilt, "The Life Cycle of Sterling Drug, Inc.," 23.

46. David R. Olmos, "German Firm to Reclaim Bayer Aspirin Name," *Los Angeles Times*, September 13, 1994, https://www.latimes.com/archives/la-xpm-1994-09-13-fi-38019-story.html.

47. Alex Scott, "Bayer: 150 Years Old and Still Inventing," *Chemical and Engineering News*, June 10, 2013, https://cen.acs.org/articles/91/i23/Bayer-150-Years-Old-Still.html.

48. Jonathan Miner and Adam Hoffhines, "The Discovery of Aspirin's Antithrombotic Effects," *Texas Heart Institute Journal* 34, no. 2 (2007): 181.

49. Miner and Hoffhines, "The Discovery of Aspirin's Antithrombotic Effects," 182.

50. H. J. Weiss and L. M. Aledort, "Impaired Platelet-Connective-Tissue Reaction in Man after Aspirin Ingestion," *Lancet* 290, no. 7514 (September 2, 1967): 495–97, doi:10.1016/s0140-6736(67)91658-3.

51. Desborough, "The Aspirin Story," 679.

52. Wood, "From Plant Extract to Molecular Panacea," 4.

53. Wood, "From Plant Extract to Molecular Panacea," 4.

54. Argentina Ornelas et al., "Beyond COX-1: The Effects of Aspirin on Platelet Biology and Potential Mechanisms of Chemoprevention," *Cancer and Metastasis Reviews* 36, no. 2 (2017): 290, doi:10.1007/s10555-017-9675-z.

55. Wood, "From Plant Extract to Molecular Panacea," 4.

56. Susan Dentzer, "Transcript: Congress Questions Vioxx, FDA," PBS, November 10, 2004, https://www.pbs.org/newshour/show/congress-questions-vioxx-fda.

57. Desborough, "The Aspirin Story," 679.

58. X. Garcia-Albeniz et al., "Aspirin for the Prevention of Colorectal Cancer," *Best Practice and Research: Clinical Gastroenterology* 25 (2011): 467–68, doi:10.1016/j.bpg.2011.10.015.

59. Garcia-Albeniz et al., "Aspirin for the Prevention of Colorectal Cancer," 461–72.

60. "Can Taking Aspirin Help Prevent Cancer?" National Institutes of Health, National Cancer Center, last updated October 7, 2020, https://www.cancer.gov/about-cancer/causes-prevention/research/aspirin-cancer-risk.

Chapter 4

1. Thomas A. Ban, "The Role of Serendipity in Drug Discovery," *Dialogues in Clinical Neuroscience* 8, no. 3 (September 2006): 337.

2. Jack F. Cade, "John Frederick Joseph Cade: Family Memories on the Occasion of the 50th Anniversary of His Discovery of the Use of Lithium in Mania," *Australian and New Zealand Journal of Psychiatry* 33, no. 5 (October 1999): 615, doi:10.1080/j.1440 -1614.1999.00624.x.

3. Phillip B. Mitchell and Dusan Hadzi-Pavlovic, "Lithium Treatment for Bipolar Disorder," *Bulletin of the World Health Organization* 78, no. 4 (2000): 515.

4. Philip B. Mitchell, "On the 50th Anniversary of John Cade's Discovery of the Anti-Manic Effect of Lithium," *Australian and New Zealand Journal of Psychiatry* 33, no. 5 (October 1999): 624, https://doi.org/10.1080/j.1440-1614.1999.00607.x.

5. Walter A. Brown, *Lithium: A Doctor, a Drug, and a Breakthrough* (New York: Norton, 2019), 63.

6. Johan Schioldann, "From Guinea Pigs to Manic Patients: Cade's 'Story of Lithium,'" *Australian and New Zealand Journal of Psychiatry* 47, no. 5 (May 2013): 48, doi:10.1177/0004867413482384.

7. Mitchell, "On the 50th Anniversary of John Cade's Discovery of the Anti-Manic Effect of Lithium," 624.

8. John F. J. Cade, "Lithium Salts in the Treatment of Psychotic Excitement," *Medical Journal of Australia* 2, no. 10 (September 3, 1949): 349.

9. Cade, "Lithium Salts in the Treatment of Psychotic Excitement," 249.

10. Mitchell, "On the 50th Anniversary of John Cade's Discovery of the Anti-Manic Effect of Lithium," 624.

11. Cade, "Lithium Salts in the Treatment of Psychotic Excitement," 350.

12. Cade, "Lithium Salts in the Treatment of Psychotic Excitement," 350–51.

13. Richard Timmer and Jeff Sands, "Lithium Intoxication," *Journal of the American Society of Nephrology* 10 (1999): 669–72.

14. Brown, *Lithium*, 91.

15. "Samuel Gershon, Lithium—Discovered, Forgotten and Rediscovered," transcript of lecture given on behalf of the History Committee of the American College of Neuropsychopharmacology meeting, December 2009, https://inhn.org/archives /gershon-collection/lithium-discovered-forgotten-and-rediscovered.html.

16. Alexander C. Kaufman, "The Original 7-Up Was a Mind-Altering Substance," *Huffington Post*, last updated December 6, 2017, https://www.huffpost.com/entry/7up -history_n_5836322.

17. Mitchell, "On the 50th Anniversary of John Cade's Discovery of the Anti-Manic Effect of Lithium," 625.

18. P. Bech, "The Full Story of Lithium," *Psychotherapy and Psychosomatics* 75 (2006): 265.

19. Mitchell, "On the 50th Anniversary of John Cade's Discovery of the Anti-Manic Effect of Lithium," 627.

20. Leonardo Tondo et al., "Clinical Use of Lithium Salts: Guide for Users and Prescribers," *International Journal of Bipolar Disorders* 22, no. 7 (July 22, 2019): 2, doi:10.1186/s40345-019-0151-2.

21. Edward Shorter, "The History of Lithium Therapy," *Bipolar Disorders* 11, no. 2, supplement (June 2009): 2, doi:10.1111/j.1399-5618.2009.00706.x.

22. Mitchell and Hadzi-Pavlovic, "Lithium Treatment for Bipolar Disorder," 516.

23. Shorter, "The History of Lithium Therapy," 2.

24. Thomas A. Ban, "The Role of Serendipity in Drug Discovery," *Dialogues in Clinical Neuroscience* 8, no. 3 (September 2006): 337.

25. William Alexander Hammond, *A Treatise on Diseases of the Nervous System* (New York: Appleton, 1881), 65, 106, 381, 582.

26. Gordon Parker, "Images in Psychiatry: John Cade," *American Journal of Psychiatry* 169, no. 2 (February 2012): 125.

27. Shorter, "The History of Lithium Therapy," 2.

28. Shorter, "The History of Lithium Therapy," 4.

29. "Database of Ionic Radii," Atomistic Simulation Group in the Materials Department of Imperial College, accessed November 2, 2020, http://abulafia.mt.ic.ac.uk/shannon/ptable.php.

30. Martin Alda, "Lithium in the Treatment of Bipolar Disorder: Pharmacology and Pharmacogenetics," *Molecular Psychiatry* 20, no. 6 (June 2015): 613, doi:10.1038/mp.2015.4.

31. Alda, "Lithium in the Treatment of Bipolar Disorder," 614.

32. Alda, "Lithium in the Treatment of Bipolar Disorder," 618.

33. Simone Pisano et al., "Putative Mechanisms of Action and Clinical Use of Lithium in Children and Adolescents: A Critical Review," *Current Neuropharmacology* 17, no. 4 (April 2019): 321–24, doi:10.2174/1570159X16666171219142120.

34. Andrea Cirpriani et al., "Lithium in the Prevention of Suicide in Mood Disorders: Updated Systematic Review and Meta-Analysis," *British Medical Journal* 346 (2013): 1, https://doi.org/10.1136/bmj.f3646.

35. Cirpriani et al., "Lithium in the Prevention of Suicide in Mood Disorders," 3.

36. M. W. Kirschner et al., "The Role of Biomedical Research in Health Care Reform," *Science* 266, no. 5182 (October 7, 1994): 49, doi:10.1126/sZience.7939643.

Chapter 5

1. Sylvia Wrobel, "Science, Serotonin, and Sadness: The Biology of Antidepressants," *FASEB Journal* 21, no. 13 (November 2007): 3408, https://doi.org/10.1096/fj.07-1102ufm.

2. Francisco López-Muñoz and Cecilio Alamo, "Monoaminergic Neurotransmission: The History of the Discovery of Antidepressants from 1950s until Today," *Current Pharmaceutical Design* 15, no. 14 (2009): 1564, doi:10.2174/138161209788168001.

3. "History of World TB Day," Centers for Disease Control and Prevention, last updated December 12, 2016, https://www.cdc.gov/tb/worldtbday/history.htm.

4. Israel Hershkovitz, "Detection and Molecular Characterization of 9000-Year-Old *Mycobacterium tuberculosis* from a Neolithic Settlement in the Eastern Mediterranean," *PLoS One* 3, no. 10 (October 2008): e3426, doi:10.1371/journal.pone.0003426.

5. H. Corwin Hinshaw, "Streptomycin in Tuberculosis," *American Journal of Medicine* 2, no. 5 (May 1, 1947): 429–35, https://doi.org/10.1016/0002-9343(47)90087-9.

6. "Fighting for Breath: Stopping the TB Epidemic," Museum of Health Care at Kingston, accessed November 5, 2020, https://www.museumofhealthcare.ca/explore/exhibits/breath/collapse-therapies.html.

7. Annika Neklason, "A Historical Lesson in Disease Containment," *Atlantic*, March 21, 2020, https://www.theatlantic.com/health/archive/2020/03/tuberculosis-sanatoriums-were-quarantine-experiment/608335.

8. López-Muñoz and Alamo, "Monoaminergic Neurotransmission," 1565.

9. Thomas A. Ban, "The Role of Serendipity in Drug Discovery," *Dialogues in Clinical Neuroscience* 8, no. 3 (September 2006): 342.

10. Wrobel, "Science, Serotonin, and Sadness," 3408.

11. Jess G. Fiedorowicz and Karen L. Swartz, "The Role of Monoamine Oxidase Inhibitors in Current Psychiatric Practice," *Journal of Psychiatric Practice* 10, no. 4 (July 2004): 241.

12. R. A. Maxwell and S. B. Eckhardt, *Drug Discovery* (Totowa, NJ: Humana Press, 1990), 146, https://doi.org/10.1007/978-1-4612-0469-5_11.

13. López-Muñoz and Alamo, "Monoaminergic Neurotransmission," 1566.

14. López-Muñoz and Alamo, "Monoaminergic Neurotransmission," 1566–67.

15. Maxwell and Eckhardt, *Drug Discovery*, 143.

16. López-Muñoz and Alamo, "Monoaminergic Neurotransmission," 1567.

17. López-Muñoz and Alamo, "Monoaminergic Neurotransmission," 1567.

18. Rebecca Kreston, "The Psychic Energizer! The Serendipitous Discovery of the First Antidepressant," *Discover*, January 27, 2016, https://www.discovermagazine.com/health/the-psychic-energizer-the-serendipitous-discovery-of-the-first-antidepressant.

19. C. M. B. Pare and M. Sandler, "A Clinical and Biochemical Study of a Trial of Iproniazid in the Treatment of Depression," *Journal of Neurology, Neurosurgery, and Psychiatry* 22, no. 3 (January 1959): 247, doi:10.1136/jnnp.22.3.247.

20. Maxwell and Eckhardt, *Drug Discovery*, 143.

21. B. K. Sinha, "Enzymatic Activation of Hydrazine Derivatives: A Spin-Trapping Study," *Journal of Biological Chemistry* 258, no. 2 (January 25, 1983): 796.

22. Wrobel, "Science, Serotonin, and Sadness," 3409.

23. López-Muñoz and Alamo, "Monoaminergic Neurotransmission," 1567.

24. Fiedorowicz and Swartz, "The Role of Monoamine Oxidase Inhibitors in Current Psychiatric Practice," 239–48.

25. "Global Health: Tuberculosis," Centers for Disease Control and Prevention, last updated April 6, 2020, https://www.cdc.gov/globalhealth/newsroom/topics/tb/index.html.

Chapter 6

1. J. Somberg, "Digitalis: Historical Development in Clinical Medicine," *Journal of Clinical Pharmacology* 25, no. 7 (1985): 484, doi:10.1177/009127008502500703.

2. M. R. Wilkins et al., "William Withering and Digitalis, 1785 to 1985," *British Medical Journal (Clinical Research Education)* 290, no. 6461 (January 5, 1985): 7, doi:10.1136/bmj.290.6461.7.

3. Cheryl Hogue, "Digoxin," *Chemical and Engineering News* 83, no. 25 (May 20, 2005), https://pubsapp.acs.org/cen/coverstory/83/8325/8325digoxin.html.

4. Wilkins et al., "William Withering and Digitalis, 1785 to 1985," 7.

5. Steven A. Edwards, "Digitalis: The Flower, the Drug, the Poison," American Association for the Advancement of Science, December 10, 2012, https://www.aaas.org/digitalis-flower-drug-poison.

6. Howard B. Burchell, "Digitalis Poisoning: Historical and Forensic Aspects," *Journal of the American College of Cardiology* 1, no. 2 (1983): 506.

7. Wilkins et al., "William Withering and Digitalis, 1785 to 1985," 7.

8. Alasdair Breckenridge, "William Withering's Legacy—For the Good of the Patient," *Clinical Medicine* 6, no. 4 (July/August 2006): 393, doi:10.7861/clinmedicine.6-4-393.

9. Hogue, "Digoxin."

10. Steven A. Edwards, "Digitalis."

11. Wilkins et al., "William Withering and Digitalis, 1785 to 1985," 7.

12. Breckenridge, "William Withering's Legacy," 393.

13. Demic Dogas and Sefik Gorkey, "Van Gogh and the Obsession of Yellow: Style or Side Effect," *Eye* 33 (2019): 165, https://doi.org/10.1038/s41433-018-0204-2.

14. Anna Gruener, "Vincent van Gogh's Yellow Vision," *British Journal of General Practice* 63, no. 612 (July 2013): 370, doi:10.3399/bjgp13X669266.

15. Dogas and Gorkey, "Van Gogh and the Obsession of Yellow," 166.

16. Dogas and Gorkey, "Van Gogh and the Obsession of Yellow," 165.

17. Gruener, "Vincent van Gogh's Yellow Vision," 370.

18. Gruener, "Vincent van Gogh's Yellow Vision," 371.

19. Sydney Smith, "Digoxin, a New Digitalis Glucoside," *Journal of the Chemical Society* (1930): 508, https://doi.org/10.1039/JR9300000508.

20. Rachel Steckelberg and James S. Newman, "The Fascinating Foxglove," *ACP Hospitalist*, March 2010, https://acphospitalist.org/archives/2010/03/newman.htm.

21. Deborah Blum, "How a Lawsuit against Coca-Cola Convinced Americans to Love Caffeine," *Time*, September 25, 2018, https://time.com/5405132/coca-cola-trial-caffeine-history.

22. Smith, "Digoxin, a New Digitalis Glucoside," 508.

23. A. Hollman et al., "Digoxin Comes from *Digitalis Lanata* and Author's Reply," *British Medical Journal* 312, no. 7035 (April 6, 1996): 912, doi:10.1136/bmj.312.7035.912.

24. Hogue, "Digoxin."

25. Hogue, "Digoxin."

26. Breckenridge, "William Withering's Legacy," 394.

27. Miguel Lopez-Lazaro, "Digoxin, HF-1, and Cancer," *Proceedings of the National Academy of Sciences* 106, no. 9 (March 3, 2009): E26, doi:10.1073/pnas.0813047106.

28. Anthony A. Bavry, "Digoxin Investigation Group—DIG," American College of Cardiology, March 11, 2013, https://www.acc.org/Latest-in-Cardiology/Clinical -Trials/2014/04/01/15/49/DIG?w_nav=Twitter.

29. Burchell, "Digitalis Poisoning," 509.

30. Charles Graeber, *The Good Nurse: A True Story of Medicine, Madness, and Murder* (New York: Hachette, 2013), 32–33.

31. Richard Pérez-Peña et al., "Death on the Night Shift: 16 Years, Dozens of Bod- ies; Through Gaps in System, Nurse Left Trail of Grief," *New York Times*, February 29, 2004, https://www.nytimes.com/2004/02/29/nyregion/death-night-shift-16-years -dozens-bodies-through-gaps-system-nurse-left-trail.html?scp=1&sq=death%20on%20 the%20night%20shift&st=cse.

32. Graeber, *The Good Nurse*, 33.

33. Pérez-Peña et al., "Death on the Night Shift."

34. Graeber, *The Good Nurse*, 6.

35. Graeber, *The Good Nurse*, 29.

36. Pérez-Peña et al., "Death on the Night Shift."

37. Charles Graeber, "The Tainted Kidney," *New York Magazine*, April 5, 2007, https://nymag.com/news/features/30331.

38. Pérez-Peña et al., "Death on the Night Shift."

39. Emily Webb, "Angels of Death Tells How Nurse Charles Cullen Killed Patients," *Herald Sun*, January 5, 2015, https://www.heraldsun.com.au/news/law-order/angels-of -death-tells-how-nurse-charles-cullen-killed-patients/news-story/3f1482705c3ee2a6ff47 238fdb1f0fa6.

40. Pérez-Peña et al., "Death on the Night Shift."

41. Pérez-Peña et al., "Death on the Night Shift."

42. Charles Graeber, "How a Serial-Killing Night Nurse Hacked Hospital Drug Protocol," *Wired*, April 29, 2013, https://www.wired.com/2013/04/charles-cullen -hospital-hack.

43. Graeber, "How a Serial-Killing Night Nurse Hacked Hospital Drug Protocol."

44. Graeber, *The Good Nurse*, 83–84.

45. Pérez-Peña et al., "Death on the Night Shift."

46. J. H. Tinker, "Sodium Nitroprusside: Pharmacology, Toxicology and Therapeu- tics," *Anesthesiology* 45, no. 3 (September 1976): 340–54.

47. Pérez-Peña et al., "Death on the Night Shift."

48. Pérez-Peña et al., "Death on the Night Shift."

49. Pérez-Peña et al., "Death on the Night Shift."

50. Graeber, "The Tainted Kidney."

51. Pérez-Peña et al., "Death on the Night Shift."

52. Graeber, "The Tainted Kidney."

53. Sophie Sohn and Allan Chernoff, "Killer Nurse Gets 11 Life Sentences," CNN, March 10, 2006, https://www.cnn.com/2006/LAW/03/02/killer.nurse.

54. Webb, "Angels of Death Tells How Nurse Charles Cullen Killed Patients."

55. Graeber, "The Tainted Kidney."

56. Graeber, "The Tainted Kidney."

57. Graeber, "The Tainted Kidney."

58. Graeber, "The Tainted Kidney."

Chapter 7

1. Thomas H. Maugh III, "Leo Sternbach, 97; Invented Valium, Many Other Drugs," *Los Angeles Times*, October 1, 2005, https://www.latimes.com/archives/la-xpm -2005-oct-01-me-sternbach1-story.html.

2. Alan Dronsfield and Pete Ellis, "Librium and Valium—Anxious Times," Royal Society of Chemistry, August 31, 2008, https://edu.rsc.org/feature/librium-and-valium -anxious-times/2020182.article.

3. Maugh, "Leo Sternbach, 97."

4. Ivan Oransky, "Leo H. Sternbach," *Lancet* 366, no. 9495 (October 22, 2005): P1430, https://doi.org/10.1016/S0140-6736(05)67588-5.

5. "Biotin—Vitamin B7," Harvard T. H. Chan School of Public Health, accessed November 19, 2020, https://www.hsph.harvard.edu/nutritionsource/biotin-vitamin-b7.

6. Werner Bonrath et al., "Biotin—The Chiral Challenge," *Chimia International Journal for Chemistry* 63, no. 5 (May 2009): 265–69, https://doi.org/10.2533/chimia.2009 .265.

7. Oransky, "Leo H. Sternbach."

8. Andrea Tone, *The Age of Anxiety: A History of America's Turbulent Affair with Tranquilizers* (New York: Basic Books, 2009), 63.

9. Leo H. Sternbach, "The Benzodiazepine Story," *Journal of Medicinal Chemistry* 22, no. 1 (January 1979): 1, https://doi.org/10.1021/jm00187a001.

10. Dronsfield and Ellis, "Librium and Valium."

11. Thomas A. Ban, "The Role of Serendipity in Drug Discovery," *Dialogues in Clinical Neuroscience* 8, no. 3 (September 2006): 340.

12. L. O. Randall et al., "Pharmacological and Clinical Studies on Valium (T.M.), a New Psychotherapeutic Agent of the Benzodiazepine Class," *Current Therapeutic Research, Clinical and Experimental* 3 (September 1961): 405–25.

13. Maugh, "Leo Sternbach, 97."

14. Maugh, "Leo Sternbach, 97."

15. Dronsfield and Ellis, "Librium and Valium."

16. "Data Sheet: Librium® C-IV (Chlordiazepoxide HCl) Capsules," U.S. Food and Drug Administration, accessed November 19, 2020, https://www.accessdata.fda.gov /drugsatfda_docs/label/2016/012249s049lbl.pdf.

17. Dronsfield and Ellis, "Librium and Valium."

18. David Hanson, "The Top Pharmaceuticals That Changed the World: Librium," *Chemical and Engineering News* 83, no. 25 (June 20, 2005), https://pubsapp.acs.org/cen /coverstory/83/8325/8325librium.html.

19. Jenny Bryan, "Landmark Drugs: The Discovery of Benzodiazepines and the Adverse Publicity That Followed," *Pharmaceutical Journal* 283 (September 18, 2009): 305.

20. Guy Chouinard, "Issues in the Clinical Use of Benzodiazepines: Potency, Withdrawal, and Rebound," *Journal of Clinical Psychiatry* 65, suppl. 5 (2004): 8.

21. "FDA Requires Strong Warnings for Opioid Analgesics, Prescription Opioid Cough Products, and Benzodiazepine Labeling Related to Serious Risks and Death from Combined Use," U.S. Food and Drug Administration, last updated August 31, 2016, https://www.fda.gov/news-events/press-announcements/fda-requires-strong-warnings-opioid-analgesics-prescription-opioid-cough-products-and-benzodiazepine.

22. Ankur Sachdeva et al., "Alcohol Withdrawal Syndrome: Benzodiazepines and Beyond," *Journal of Clinical and Diagnostic Research* 9, no. 9 (September 2015): VE01–07, doi:10.7860/JCDR/2015/13407.6538.

23. Oransky, "Leo H. Sternbach."

24. Dronsfield and Ellis, "Librium and Valium."

25. Oransky, "Leo H. Sternbach."

26. Maugh, "Leo Sternbach, 97."

27. Oransky, "Leo H. Sternbach."

28. Maugh, "Leo Sternbach, 97."

Chapter 8

1. J. R. Partington, *A History of Chemistry* (London: Palgrave, 1962), 237.

2. James Marshall and Virginia Marshall, "Rediscovery of the Elements: Joseph Priestly," *The Hexagon* 96, no. 2 (Summer 2005): 29–31.

3. V. Lew et al., "Past, Present, and Future of Nitrous Oxide," *British Medical Bulletin* 125, no. 1 (March 2018): 103, https://doi.org/10.1093/bmb/ldx050.

4. Peter Ford, "Joseph Priestly: The Man Who Discovered Oxygen," Bath Royal Literary and Scientific Institution, January 2004, https://www.brlsi.org/joseph-priestley-man-who-discovered-oxygen.

5. Ford, "Joseph Priestly."

6. Marshall and Marshall, "Rediscovery of the Elements," 30.

7. F. F. Cartwright, "Humphry Davy's Researches on Nitrous Oxide," *British Journal of Anaesthesia* 44 (1972): 293.

8. Marshall and Marshall, "Rediscovery of the Elements," 31.

9. "Joseph Priestly," Science History Institute, last modified December 14, 2017, https://www.sciencehistory.org/historical-profile/joseph-priestley.

10. Marshall and Marshall, "Rediscovery of the Elements," 31.

11. Marshall and Marshall, "Rediscovery of the Elements," 33.

12. Nicholas Riegels et al., "Humphry Davy: His Life, Works, and Contribution to Anesthesiology," *Anesthesiology* 114, no. 6 (June 2011): 1282.

13. Cartwright, "Humphry Davy's Researches on Nitrous Oxide," 291.

14. Cartwright, "Humphry Davy's Researches on Nitrous Oxide," 292.

15. Riegels et al., "Humphry Davy," 1283.

16. Guillaume Moulis and Guillaume Martin-Blondel, "Scrofula, the King's Evil," *Canadian Medical Association Journal* 184, no. 9 (June 12, 2012): 1061, doi:10.1503/cmaj.111519.

17. Cartwright, "Humphry Davy's Researches on Nitrous Oxide," 291.

18. Cartwright, "Humphry Davy's Researches on Nitrous Oxide," 294.

19. Cartwright, "Humphry Davy's Researches on Nitrous Oxide," 294–96.

20. Humphry Davy, "On a New Detonating Compound, in a Letter from Sir Humphry Davy, LL. D. F. R.S. to the Right Honourable Sir Joseph Banks, Bart. K. B. P. R. S," *Philosophical Transactions of the Royal Society* 103 (December 31, 1813): 1–7, http://doi.org/10.1098/rstl.1813.0002.

21. Riegels et al., "Humphry Davy," 1286.

22. Riegels et al., "Humphry Davy," 1287.

23. Mark Crawford, "Samuel Colt," American Society of Mechanical Engineers, April 18, 2012, https://www.asme.org/topics-resources/content/samuel-colt.

24. Henry Wood Erving, "The Discoverer of Anaesthesia: Dr. Horace Wells of Hartford," *Yale Journal of Biology and Medicine* 5, no. 5 (May 1933): 424–25.

25. Erving, "The Discoverer of Anaesthesia," 425.

26. Erving, "The Discoverer of Anaesthesia," 426.

27. Katherine Liu et al., "Last Days of Horace Wells: A Sad Story of His Arrest and a Suicide," American Society of Anesthesiologists annual meeting, October 13, 2014, http://www.asaabstracts.com/strands/asaabstracts/abstract.htm?year=2014&index=12&absnum=2419.

28. Rajesh P. Haridas, "'Gentlemen! This Is No Humbug': Did John Collins Warren, M.D., Proclaim These Words on October 16, 1846, at Massachusetts General Hospital, Boston?" *Anesthesiology* 124, no. 3 (March 2016): 553–60, https://doi.org/10.1097/ALN.0000000000000944.

29. Erving, "The Discoverer of Anaesthesia," 428–29.

30. Lew et al., "Past, Present, and Future of Nitrous Oxide," 104.

31. "A Strange Story, Facts Briefly States. The Colton Dental Association—A Word from the Medical Profession," *New York Times*, March 24, 1865, https://www.nytimes.com/1865/03/25/archives/a-strange-story-facts-briefly-stated-the-colton-dental-association.html.

32. David Zuck, "Nitrous Oxide: Are You Having a Laugh?" Royal Society of Chemistry, February 29, 2012, https://edu.rsc.org/feature/nitrous-oxide-are-you-having-a-laugh/2020202.article.

33. Lew et al., "Past, Present, and Future of Nitrous Oxide," 105.

34. Lew et al., "Past, Present, and Future of Nitrous Oxide," 110–11.

35. Dimitris E. Emmanoui and Raymond M. Quock, "Advances in Understanding the Actions of Nitrous Oxide," *Anesthesia Progress* 54, no. 1 (Spring 2007): 9–18.

36. Lew et al., "Past, Present, and Future of Nitrous Oxide," 111.

37. G. Randhawa and A. Bodenham, "The Increasing Recreational Use of Nitrous Oxide: History Revisited," *British Journal of Anaesthesia* 116, no. 3 (August 30, 2015): 321–24.

Chapter 9

1. Johnathan Frunzi, "From Weapon to Wonder Drug," *Hospitalist*, February 2007, https://www.the-hospitalist.org/hospitalist/article/123282/weapon-wonder-drug.

2. "Germans Introduce Poison Gas," History, February 9, 2010, https://www.history.com/this-day-in-history/germans-introduce-poison-gas.

3. Gerard J. Fitzgerald, "Chemical Warfare and Medical Response during World War I," *American Journal of Public Health* 98, no. 4 (April 2008): 612, doi:10.2105/AJPH.2007.11930.

4. Michael Tinnesand, "Mustard Gas," *ChemMatters*, April 18, 2005.

5. "Mapping Chaucer: Flanders," Mapping Chaucer, accessed December 8, 2020, https://mediakron.bc.edu/mappingchaucer/the-canterbury-tales-1/flanders-4.

6. Fitzgerald, "Chemical Warfare and Medical Response during World War I," 611–12.

7. Fitzgerald, "Chemical Warfare and Medical Response during World War I," 617.

8. "Fritz Haber's Synthesis of Ammonia from Its Elements, Hydrogen and Nitrogen, Earned Him the 1918 Nobel Prize in Chemistry," Science History Institute, last updated December 7, 2017, https://www.sciencehistory.org/historical-profile/fritz-haber.

9. Fitzgerald, "Chemical Warfare and Medical Response during World War I," 615.

10. Fitzgerald, "Chemical Warfare and Medical Response during World War I," 617.

11. A. P. Padley, "Gas: The Greatest Terror of the Great War," *Anaesthesia and Intensive Care* 44, supplement (July 2016): 29, doi:10.1177/0310057X1604401S05.

12. Padley, "Gas," 28.

13. Fitzgerald, "Chemical Warfare and Medical Response during World War I," 618.

14. Tinnesand, "Mustard Gas," 18.

15. "Germans Introduce Poison Gas."

16. Fitzgerald, "Chemical Warfare and Medical Response during World War I," 612.

17. Tinnesand, "Mustard Gas," 17.

18. Tinnesand, "Mustard Gas," 19.

19. Carolyn Wilke, "From Chemical Weapon to Chemotherapy, 1917–1946," *Scientist*, March 31, 2019, https://www.the-scientist.com/foundations/from-chemical-weapon-to-chemotherapy--19171946-65655.

20. Jennet Conant, "How a WWII Disaster—and Cover-Up—Led to a Cancer Treatment Breakthrough," History, August 12, 2020, https://www.history.com/news/wwii-disaster-bari-mustard-gas.

21. Jerzy Einhorn, "Nitrogen Mustard: The Origin of Chemotherapy for Cancer," *International Journal of Radiation Oncology, Biology, Physics* 11, no. 7 (July 1985): 1375–78, doi:10.1016/0360-3016(85)90254-8.

22. Einhorn, "Nitrogen Mustard," 1375–78.

23. "Emergency Preparedness and Response: Facts about Nitrogen Mustards," Centers for Disease Control and Prevention, last updated April 4, 2018, https://emergency.cdc.gov/agent/nitrogenmustard/index.asp.

24. "Cancer: A Historic Perspective," National Institutes of Health National Cancer Institute, accessed December 7, 2020, https://training.seer.cancer.gov/disease/history.

25. Frunzi, "From Weapon to Wonder Drug."

26. John Curtis, "From the Field of Battle, an Early Strike at Cancer," *Yale Medicine Magazine*, Summer 2005, 16.

27. Panos Christakis, "The Birth of Chemotherapy at Yale Bicentennial Lecture Series: Surgery Grand Round," *Yale Journal of Biology and Medicine* 84, no. 2 (June 2011): 170.

28. Curtis, "From the Field of Battle, an Early Strike at Cancer," 17.

29. Christakis, "The Birth of Chemotherapy at Yale Bicentennial Lecture Series," 171.

30. Curtis, "From the Field of Battle, an Early Strike at Cancer," 17.

31. Tinnesand, "Mustard Gas," 19.

32. Ali Karmai and Shahriar Khateri, "Long Legacy," *CBRNe World*, August 2012, 38.

33. "Remembering the Halabja Massacre," Voice of America, March 15, 2018, https://editorials.voa.gov/a/remembering-halabja-massacre/4298678.html.

34. "Mechlorethamine," U.S. National Library of Medicine: MedlinePlus, last updated August 15, 2012, https://medlineplus.gov/druginfo/meds/a682223.html.

35. D. A. Karnofsky et al., "The Use of the Nitrogen Mustards in the Palliative Treatment of Carcinoma: With Particular Reference to Bronchogenic Carcinoma," *Cancer* 1, no. 4 (November 1948): 635.

36. Y. H. Kim et al., "Topical Nitrogen Mustard in the Management of Mycosis Fungoides: Update of the Stanford Experience," *Archives of Dermatology* 139, no. 2 (February 2003): 165–66, doi:10.1001/archderm.139.2.165.

37. Laurence L. Brunton et al. *Goodman and Gilman's the Pharmacological Basis of Therapeutics*, 11th ed. (New York: McGraw-Hill, 2006), 1322–23.

Chapter 10

1. Charles W. Francis, "Warfarin: An Historical Perspective," *Hematology*, no. 1 (2008): 251, https://doi.org/10.1182/asheducation-2008.1.251.

2. Frank W. Schofield, "A Brief Account of a Disease in Cattle Simulating Hemorrhagic Septicemia Due to Feeding Sweet Clover," *Canadian Veterinary Journal* 25, no. 12 (1984): 453–55.

3. Robert H. Burris, "8. Karl Paul Link," *National Academy of Sciences Biographical Memoirs 65* (1994): 177–79, doi:10.17226/4548.

4. Burris, "8. Karl Paul Link," 180–81.

5. B. M. Duxbury and Leon Poller, "The Oral Anticoagulant Saga: Past, Present, and Future," *Clinical and Applied Thrombosis/Hemostasis* 7, no. 4 (October 2001): 270, doi:10.1177/107602960100700403.

6. Duxbury and Poller, "The Oral Anticoagulant Saga," 270.

7. Karl P. Link, "The Discovery of Dicumarol and Its Sequels," *Circulation* 19, no. 1 (January 1959): 99, doi:10.1161/01.cir.19.1.97.

8. Gregory B. Lim, "Warfarin: From Rat Poison to Clinical Use," *Nature Reviews Cardiology*, December 14, 2017, https://doi.org/10.1038/nrcardio.2017.172.

9. Link, "The Discovery of Dicumarol and Its Sequels," 101.

10. Burris, "8. Karl Paul Link," 187.

11. Lim, "Warfarin."

12. Duxbury and Poller, "The Oral Anticoagulant Saga," 270.

13. Link, "The Discovery of Dicumarol and Its Sequels," 103.

14. Lim, "Warfarin."

15. Link, "The Discovery of Dicumarol and Its Sequels," 104.

16. Burris, "8. Karl Paul Link," 188.

17. Burris, "8. Karl Paul Link," 188.

18. Link, "The Discovery of Dicumarol and Its Sequels," 97–107.

19. Burris, "8. Karl Paul Link," 188–89.

20. Burris, "8. Karl Paul Link," 189–92.

21. Link, "The Discovery of Dicumarol and Its Sequels," 104.

22. Ramya Rajagopalan, "A Study in Scarlet," Science History Institute, March 29, 2018, https://www.sciencehistory.org/distillations/a-study-in-scarlet.

23. "They Built a Better Mousetrap . . . and Used Radio to Sell It," *Broadcasting Telecasting* 39, no. 24 (December 11, 1950): 22–23, https://worldradiohistory.com/Archive-BC/BC-1950/BC-1950-12-11.pdf.

24. Rajagopalan, "A Study in Scarlet."

25. Link, "The Discovery of Dicumarol and Its Sequels," 105.

26. Lim, "Warfarin."

27. Link, "The Discovery of Dicumarol and Its Sequels," 106.

28. Laurence L. Brunton et al., *Goodman and Gilman's the Pharmacological Basis of Therapeutics*, 11th ed. (New York: McGraw-Hill, 2006), 1476–77.

29. Lim, "Warfarin."

30. Munir Pirmohamed, "Warfarin: Almost 60 Years Old and Still Causing Problems," *British Journal of Clinical Pharmacology* 62, no. 5 (November 2006): 509, doi:10.1111/j.1365-2125.2006.02806.x.

31. Pirmohamed, "Warfarin," 510.

32. Pirmohamed, "Warfarin," 510.

33. Pirmohamed, "Warfarin," 509.

34. "Warfarin and Vitamin K," Healthlink British Columbia, accessed December 3, 2021, https://www.healthlinkbc.ca/healthy-eating-physical-activity/food-and-nutrition/nutrients/food-sources-vitamin-k.

35. Michael Wines, "New Study Supports Idea Stalin Was Poisoned," *New York Times*, March 5, 2003.

36. Nicholas Johnson et al, "Vampire Bat Rabies: Ecology, Epidemiology and Control," *Viruses* 6, no. 5 (May 2014): 1911–28, doi:10.3390/v6051911.

37. Hans-Joachim Pelz et al., "The Genetic Basis of Resistance to Anticoagulants in Rodents," *Genetics* 170, no. 4 (August 2005): 1839–47, doi:10.1534/genetics.104.040360.

Chapter 11

1. Frank J. Ergbuth, "Historical Notes on Botulism, Clostridium Botulinum, Botulinum Toxin, and the Idea of the Therapeutic Use of the Toxin," *Movement Disorders* 18, suppl. 8 (March 2004): S2, doi:10.1002/mds.20003.

2. "Pruno: A Recipe for Botulism," Centers for Disease Control, October 4, 2018, https://www.cdc.gov/botulism/pruno-a-recipe-for-botulism.html.

3. Vishwanath S. Hanchanale et al., "The Unusual History and the Urological Applications of Botulinum Neurotoxin," *Urologia Internationalis* 85, no. 2 (September 2010): 126, doi:10.1159/000317517.

4. J. Jankovic, "Botulinum Toxin in Clinical Practice," *Journal of Neurology, Neurosurgery, and Psychiatry* 75, no. 7 (July 2004): 951, doi:10.1136/jnnp.2003.034702.

5. Ergbuth, "Historical Notes on Botulism, Clostridium Botulinum, Botulinum Toxin, and the Idea of the Therapeutic Use of the Toxin," S2.

6. Jankovic, "Botulinum Toxin in Clinical Practice," 951.

7. "Botulism," World Health Organization, last updated January 10, 2018, https://www.who.int/news-room/fact-sheets/detail/botulism.

8. "Botulism."

9. Jankovic, "Botulinum Toxin in Clinical Practice," 953.

10. Alan B. Scott, "Botulinum Toxin Injection into Extraocular Muscles as an Alternative to Strabismus Surgery," *Ophthalmology* 87, no. 10 (October 1980): 1044–49, doi:10.1016/s0161-6420(80)35127-0.

11. Alan B. Scott, "Botulinum Toxin Injection of Eye Muscles to Correct Strabismus," *Transactions of the American Ophthalmological Society* 79 (1981): 734–70.

12. Chuck McCutcheon, "The Creator of Botox Never Cared about Wrinkles," *Scientific American*, November 3, 2016, https://blogs.scientificamerican.com/guest-blog/the-creator-of-botox-never-cared-about-wrinkles.

13. Vauhini Vara, "Billions and Billions for Botox," *New Yorker*, November 18, 2014, https://www.newyorker.com/business/currency/actavis-allergan-botox-sale.

14. Jane L. Halpern and William H. Habig, "An Overview of Some Issues in the Licensing of Botulinum Toxins," in *Botulinum and Tetanus Neurotoxins*, ed. Bibhuti R. DasGupta (Boston: Springer, 1993), 665.

15. Richard P. Clark, "The First Cosmetic Use of Botulinum Toxin," *Plastic and Reconstructive Surgery* 144, no. 4 (October 2019): 723e–24e, doi:10.1097/PRS.0000000000006081.

16. Anne Kingston, "Meet the Vancouver Couple Who Pioneered Botox," *Maclean's*, June 3, 2014, https://www.macleans.ca/society/health/meet-the-vancouver-couple-who-pioneered-botox.

17. Jean Carruthers and Alastair Carruthers, "Practical Cosmetic Botox Techniques," *Journal of Cutaneous Medicine and Surgery* 3, suppl. 4 (December 1999): S49.

18. Eric S. Felber, "Botulinum Toxin in Primary Care Medicine," *Journal of the American Osteopathic Association* 106, no. 10 (October 2006): 610.

19. Deborah Kotz, "FDA Approves Botox Injections to Treat 'Crow's Feet,'" *Boston Globe*, September 16, 2013, https://www.bostonglobe.com/lifestyle/health

-wellness/2013/09/15/fda-approves-botox-injections-treat-crow-feet-lines-around-eyes/aQG8kpWoVpC4YaAaAuZNnJ/story.html.

20. Vara, "Billions and Billions for Botox."

21. Tomi Kilgore, "AbbVie Confirms Deal to Buy Allergan for 45% Premium, Valuing Allergan at $61.7 Billion," MarketWatch, June 25, 2019, https://www.marketwatch.com/story/abbvie-confirms-deal-to-buy-allergan-for-45-premium-valuing-allergan-at-617-billion-2019-06-25.

22. Jankovic, "Botulinum Toxin in Clinical Practice," 952.

23. "Medication Guide: Myoblock (RimabotulinumtoxinB) Injection," U.S. Food and Drug Administration, 2009, https://www.fda.gov/media/77367/download.

24. Roshni Ramachandran and Tony L. Yaksh, "Therapeutic use of Botulinum Toxin in Migraine: Mechanisms of Action," *British Journal of Pharmacology* 171, no. 18 (September 2014): 4178, doi:10.1111/bph.12763.

25. Tom Watkins, "FDA Approves Botox as Migraine Preventive," CNN, October 15, 2010, http://www.cnn.com/2010/HEALTH/10/15/migraines.botox/index.html.

26. Ramachandran and Yaksh, "Therapeutic Use of Botulinum Toxin in Migraine," 4178.

27. Eric S. Felber, "Botulinum Toxin in Primary Care Medicine," *Journal of the American Osteopathic Association* 106, no. 10 (October 2006): 610–11.

28. Felber, "Botulinum Toxin in Primary Care Medicine," 610–12.

29. Jankovic, "Botulinum Toxin in Clinical Practice," 955.

30. Bat-Chen Friedman and Ran D. Goldman, "Use of Botulinum Toxin A in Management of Children with Cerebral Palsy," *Canadian Family Physician* 57, no. 9 (September 2011): 1006–7.

31. "2019 Plastic Surgery Statistics Report," American Society of Plastic Surgeons National Clearinghouse of Plastic Surgery Procedural Statistics, 2020, https://www.plasticsurgery.org/documents/News/Statistics/2019/plastic-surgery-statistics-full-report-2019.pdf.

32. "Pharmacoeconomic Review Report for Dysport Therapeutic," *CADH Common Drug Review*, August 2017, https://www.ncbi.nlm.nih.gov/books/NBK535069/pdf/Bookshelf_NBK535069.pdf.

33. Erica Simons, "Faith, Fanaticism, and Fear: Aum Shinrikyo—The Birth and Death of a Terrorist Organization," *Forensic Examiner* 15, no. 1 (Spring 2006): 37–45.

34. "Aum Shinrikyo: The Japanese Cult behind the Tokyo Sarin Attack," BBC News, July 6, 2018, https://www.bbc.com/news/world-asia-35975069.

35. Kyle B. Olson, "Aum Shinrikyo: Once and Future Threat?" *Emerging Infectious Diseases* 5, no. 4 (August 1999): 413–16, doi:10.3201/eid0504.990409.

36. Tim Parsons, "Johns Hopkins Working Group on Civilian Biodefense Says Botulinum Toxin Is a Major Biological Weapons Threat," Johns Hopkins Bloomberg School of Public Health, February 28, 2001, https://publichealth.jhu.edu/2001/botulinum-toxin-release-2001.

37. Eric S. Felber, "Botulinum Toxin in Primary Care Medicine," *Journal of the American Osteopathic Association* 106, no. 10 (October 2006): 609.

38. Michael Stebbins, "Biosecurity and Biodefense Resource: Botulinum Toxin Fact Sheet," Federation of American Scientists, accessed February 5, 2021, https://fas.org/biosecurity/resource/factsheets/botulinum.htm.

39. Peter J. Osterbauer and Michael R. Dobbs, "Neurobiological Weapons," in *Clinical Neurotoxicology*, ed. Michael R. Dobbs (Philadelphia: Saunders, 2009), 631.

40. John G. Sotos, "Botulinum Toxin in Biowarfare," *Journal of the America Medical Association* 21, no. 285 (2001): 2716, doi:10.1001/jama.285.21.2716.

41. Osterbauer and Dobbs, "Neurobiological Weapons," 631.

42. Duc J. Viglia et al., "Botulism from Drinking Pruno," *Emerging Infectious Diseases* 15, no. 1 (January 2009): 69.

43. Viglia et al., "Botulism from Drinking Pruno," 70.

44. Scott Hensley, "Botulism from 'Pruno' Hits Arizona Prison," NPR, February 7, 2013, https://www.npr.org/sections/health-shots/2013/02/07/171385104/botulism-from-pruno-hits-arizona-prison.

45. Diana Thurston et al., "Botulism from Drinking Prison-Made Illicit Alcohol—Utah 2011," *Morbidity and Mortality Weekly Report* 61, no. 39 (October 5, 2012): 782–84, https://www.cdc.gov/mmwr/preview/mmwrhtml/mm6139a2.htm.

46. Corey M. Peak et al., "Wound Botulism Outbreak among Persons Who Use Black Tar Heroin—San Diego County, California, 2017–2018," *Centers for Disease Control Weekly* 67, no. 5152 (January 4, 2019): 1415–18, https://www.cdc.gov/mmwr/volumes/67/wr/mm675152a3.htm.

47. "Botulism."

48. Elida E. Vanella de Cuetos et al., "Equine Botulinum Antitoxin for the Treatment of Infant Botulism," *Clinical and Vaccine Immunology* 18, no. 11 (November 2011): 1846, doi:10.1128/CVI.05261-11.

49. Vanella de Cuetos et al., "Equine Botulinum Antitoxin for the Treatment of Infant Botulism," 1845–49.

Chapter 12

1. Judith H. J. Roelofzen et al., "No Increased Risk of Cancer after Coal Tar Treatment in Patients with Psoriasis or Eczema," *Journal of Investigational Dermatology* 130, no. 4 (April 2010): 953, doi:10.1038/jid.2009.389.

2. Roelofzen et al., "No Increased Risk of Cancer after Coal Tar Treatment in Patients with Psoriasis or Eczema," 953.

3. "Coal to Make Coke and Steel," University of Kentucky Geological Survey, accessed February 18, 2020, https://www.uky.edu/KGS/coal/coal-for-cokesteel.php.

4. Walter Sneader, *Drug Discovery: A History* (New York: Wiley, 2005), 356.

5. "Wright's Coal Tar Soap," Grace's Guide to British Industrial History, accessed February 18, 2020, https://www.gracesguide.co.uk/Wright%27s_Coal_Tar_Soap.

6. "William Valentine Wright, Senior," Grace's Guide to British Industrial History, accessed February 11, 2020, https://www.gracesguide.co.uk/William_Valentine_Wright,_Senior.

7. B. Eriksson et al., "Erysipelas: Clinical and Bacteriologic Spectrum and Serological Aspects," *Clinical Infectious Diseases* 23, no. 5 (November 1996): 1091, doi:10.1093/clinids/23.5.1091.

8. "Wright's Coal Tar Soap."

9. Samuel Spratly, "Scabies Successfully Treated with Coal-Tar Naphtha," *British Medical Journal* 2, no. 44 (November 2, 1861): 465.

10. David MacDonald, "Antiseptic Coal-Tar Dyes," *British Medical Journal Supplement* 1, no. 3199 (April 22, 1922): 662.

11. S. A. Ashmore and A. W. McKenny Hughes, "Coal-Tar Naphtha for Destruction of Bed-Bugs," *British Medical Journal* 1, no. 4020 (January 22, 1938): 160–64, doi:10.1136/bmj.1.4020.160.

12. Trevor Brown, Alan Dronsfield, and Peter Ellis, "Pain Relief: From Coal Tar to Paracetamol," Royal Society of Chemistry, June 30, 2005, https://edu.rsc.org/feature/pain-relief-from-coal-tar-to-paracetamol/2020140.article.

13. S. Ranganathan and T. Mukhopadhyay, "Dandruff: The Most Commercially Exploited Skin Disease," *Indian Journal of Dermatology* 55, no. 2 (April–June 2010): 130, doi:10.4103/0019-5154.62734.

14. "Coal Tar Shampoo—Topical," British Columbia HealthLinkBC, last updated June 2018, https://www.healthlinkbc.ca/medications/fdb7226.

15. Ellen H. van den Bogaard et al., "Coal Tar Induces AHR-Dependent Skin Barrier Repair in Atopic Dermatitis," *Journal of Clinical Investigation* 123, no. 2 (January 25, 2013): 917, https://doi.org/10.1172/JCI65642.

16. Joshia A. Zeichner, "Use of Topical Coal Tar Foam for the Treatment of Psoriasis in Difficult-to-Treat Areas," *Journal of Clinical and Aesthetic Dermatology* 3, no. 9 (September 2010): 39.

17. "Occupational Exposures in Coal-Tar Distillation," in *Chemical Agents and Related Occupations: Volume 100 F. A Review of Human Carcinogens* (Lyon: World Health Organization, 2012), 153.

18. "Occupational Exposures in Coal-Tar Distillation," 155.

19. I. Berenblum, "Liquor Picis Carbonis (B.P.)—A Carcinogenic Agent," *British Medical Journal* 2, no. 4577 (September 24, 1948): 601, doi:10.1136/bmj.2.4577.601.

20. "Coal Tar Products Target of California Lawsuit Alleging Cancer Risk," National Psoriasis Foundation, March 22, 2001, https://web.archive.org/web/20021026015030/http://www.psoriasis.org/coaltar.htm.

21. "Subpart H—Drug Products for the Control of Dandruff, Seborrheic Dermatitis, and Psoriasis," Code of Federal Regulations Title 21, last updated November 10, 2020, https://www.accessdata.fda.gov/scripts/cdrh/cfdocs/cfcfr/CFRSearch.cfm?CFRPart=358&showFR=1&subpartNode=21:5.0.1.1.30.8.

22. "Section 740.18 Coal Tar Hair Dyes Posing a Risk of Cancer," Code of Federal Regulations Title 21, last updated November 10, 2020, https://www.accessdata.fda.gov/scripts/cdrh/cfdocs/cfcfr/CFRSearch.cfm?fr=740.18.

23. Roelofzen et al., "No Increased Risk of Cancer after Coal Tar Treatment in Patients with Psoriasis or Eczema," 953.

Chapter 13

1. J. Bryan, "How Minoxidil Was Transformed from an Antihypertensive to Hair-Loss Drug," *Pharmaceutical Journal*, July 20, 2011, https://pharmaceutical-journal.com/news-and-analysis/how-minoxidil-was-transformed-from-an-antihypertensive-to-hair-loss-drug/11080942.article.

2. Bryan, "How Minoxidil Was Transformed from an Antihypertensive to Hair-Loss Drug."

3. Bryan, "How Minoxidil Was Transformed from an Antihypertensive to Hair-Loss Drug."

4. Bryan, "How Minoxidil Was Transformed from an Antihypertensive to Hair-Loss Drug."

5. Ajaj Raj, "How a Doctor's Chance Appointment with a Hairy Woman Led to the Discovery of Rogaine," *Business Insider*, September 26, 2014, https://www.businessinsider.com/hairy-woman-led-guinter-kahn-to-rogaine-2014-9.

6. Raj, "How a Doctor's Chance Appointment with a Hairy Woman Led to the Discovery of Rogaine."

7. Raj, "How a Doctor's Chance Appointment with a Hairy Woman Led to the Discovery of Rogaine."

8. Howard Cohen, "Guinter Kahn, Inventor of Hair Loss Remedy Rogaine, Dies at 80 in Miami Beach," *Miami Herald*, September 24, 2014, https://www.miamiherald.com/news/local/obituaries/article2227194.html.

9. Dahlia Saleh et al., "Hypertrichosis," National Center for Biotechnology Information StatPearls, last updated January 5, 2021, https://www.ncbi.nlm.nih.gov/books/NBK534854.

10. Raj, "How a Doctor's Chance Appointment with a Hairy Woman Led to the Discovery of Rogaine."

11. Bryan, "How Minoxidil Was Transformed from an Antihypertensive to Hair-Loss Drug."

12. David A. Fenton and John D. Wilkinson, "Topical Minoxidil in the Treatment of Alopecia Areata," *British Medical Journal* 287, no. 6398 (October 8, 1983): 1018–20, doi:10.1136/bmj.287.6398.1015.

13. Bryan, "How Minoxidil Was Transformed from an Antihypertensive to Hair-Loss Drug."

14. Gersh Kuntzman, *Hair! Mankind's Historic Quest to End Baldness* (New York: Random House, 2001), 123.

15. Associated Press, "Hair-Growth Drug to Be Sold over the Counter," *New York Times*, February 13, 1996, https://www.nytimes.com/1996/02/13/science/hair-growth-drug-to-be-sold-over-the-counter.html.

16. Talel Badri et al., "Minoxidil," National Center for Biotechnology Information StatPearls, May 4, 2020, https://www.ncbi.nlm.nih.gov/books/NBK482378.

17. Badri et al., "Minoxidil."

18. Poonkiat Suchonwanit et al., "Minoxidil and Its Use in Hair Disorders: A Review," *Drug Design, Development, and Therapy* 13 (August 9, 2019): 2777, doi:10.2147/DDDT.S214907.

19. Suchonwanit et al., "Minoxidil and Its Use in Hair Disorders," 2779.

20. Cohen, "Guinter Kahn, Inventor of Hair Loss Remedy Rogaine, Dies at 80 in Miami Beach."

21. Cohen, "Guinter Kahn, Inventor of Hair Loss Remedy Rogaine, Dies at 80 in Miami Beach."

22. Suchonwanit et al., "Minoxidil and Its Use in Hair Disorders," 2778.

23. Suchonwanit et al., "Minoxidil and Its Use in Hair Disorders," 2778.

24. Fenton and Wilkinson, "Topical Minoxidil in the Treatment of Alopecia Areata," 1018.

25. Suchonwanit et al., "Minoxidil and Its Use in Hair Disorders," 2778.

26. "Treating Female Pattern Hair Loss," Harvard Health Publishing via Harvard Medical School, last updated August 31, 2020, https://www.health.harvard.edu/staying-healthy/treating-female-pattern-hair-loss.

27. Amanda MacMillan, "A Popular Hair Loss Drug Costs 40% More for Women Than Men," *Time*, June 9, 2017, https://time.com/4811985/rogaine-women-hair-loss-men.

28. Mackenzie R. Wehner et al., "Association between Gender and Drug Cost for Over-the-Counter Minoxidil," *JAMA Dermatology* 153, no. 8 (August 2017): 825–26, doi:10.1001/jamadermatol.2017.1394.

29. "Efficacy and Safety of 3% Minoxidil Lotion for Chest Hair Enhancement," U.S. National Library of Medicine, last updated December 2, 2014, https://clinicaltrials.gov/ct2/show/NCT02283645.

30. Suchonwanit et al., "Minoxidil and Its Use in Hair Disorders," 2780.

31. Suchonwanit et al., "Minoxidil and Its Use in Hair Disorders," 2782.

32. Suchonwanit et al., "Minoxidil and Its Use in Hair Disorders," 2782.

33. Camille DeClementi et al., "Suspected Toxicosis after Topical Administration of Minoxidil in 2 Cats," *Journal of Veterinary Emergency and Critical Care* 14, no. 4 (November 24, 2004): 287, https://doi.org/10.1111/j.1476-4431.2004.04014.x.

Chapter 14

1. Milt Freudenheim, "Keeping the Pipeline Filled at Merck," *New York Times*, February 16, 1992, https://www.nytimes.com/1992/02/16/business/keeping-the-pipeline-filled-at-merck.html?src=pm.

2. "The Extraordinary Case of the Guevedoces," BBC News, September 20, 2015, https://www.bbc.com/news/magazine-34290981.

3. Freudenheim, "Keeping the Pipeline Filled at Merck."

4. "Prostate Enlargement (Benign Prostatic Hyperplasia)," National Institutes of Diabetes and Digestive and Kidney Diseases, last updated September 2014, https://

www.niddk.nih.gov/health-information/urologic-diseases/prostate-problems/prostate
-enlargement-benign-prostatic-hyperplasia.

5. NCI Staff, "Prostate Cancer Prevention and Finasteride: A Conversation with NCI's Dr. Howard Parnes," National Cancer Institute, May 13, 2019, https://www.cancer.gov/news-events/cancer-currents-blog/2019/prostate-cancer-prevention-finasteride-parnes.

6. Patrick M. Zito et al., "Finasteride," National Center for Biotechnology Information StatPearls, last updated October 27, 2020, https://www.ncbi.nlm.nih.gov/books/NBK513329.

7. Christiane Anne Ganzer et al., "Emotional Consequences of Finasteride: Fool's Gold," *American Journal of Men's Health* 12, no. 1 (January 2018): 91, doi:10.1177/1557988316631624.

8. K. J. McClellan and A. Markham, "Finasteride: A Review of Its Use in Male Pattern Hair Loss," *Drugs* 57, no. 1 (January 1999): 111–26, doi:10.2165/00003495-19957010-00014.

9. Dan Levine and Chad Terhune, "Merck Anti-Baldness Drug Propecia Has Long Trail of Suicide Reports, Records Show," Reuters, February 3, 2021, https://www.reuters.com/article/us-merck-propecia-suicide-exclusive/exclusive-merck-anti-baldness-drug-propecia-has-long-trail-of-suicide-reports-records-show-idUSKBN2A32XU.

10. Dan Levine, "Court Let Merck Hide Secrets about a Popular Drug's Risks," Reuters, September 11, 2019, https://www.reuters.com/investigates/special-report/usa-courts-secrecy-propecia.

11. Goren Andy et al., "Controversies in the Treatment of Androgenetic Alopecia: The History of Finasteride," *Dermatologic Therapy* 32, no. 2 (March 2019): e12647, doi:10.1111/dth.12647.

12. Sung Won Lee et al., "A Systematic Review of Topical Finasteride in the Treatment of Androgenetic Alopecia in Men and Women," *Journal of Drugs in Dermatology* 17, no. 4 (April 1, 2018): 2.

13. Zito et al., "Finasteride."

14. "FDA Pregnancy Categories," U.S. Department of Health and Human Services Chemical Hazards and Emergency Management, accessed March 11, 2021, https://chemm.nlm.nih.gov/pregnancycategories.htm.

15. "Propecia Prescribing Information," U.S. Food and Drug Administration, accessed March 11, 2021, https://www.accessdata.fda.gov/drugsatfda_docs/label/2012/020788s020s021s023lbl.pdf.

16. Lee et al., "A Systematic Review of Topical Finasteride in the Treatment of Androgenetic Alopecia in Men and Women," 3.

17. Lee et al., "A Systematic Review of Topical Finasteride in the Treatment of Androgenetic Alopecia in Men and Women," 4.

18. NCI Staff, "Prostate Cancer Prevention and Finasteride: A Conversation with NCI's Dr. Howard Parnes," National Cancer Institute, May 13, 2019, https://www.cancer.gov/news-events/cancer-currents-blog/2019/prostate-cancer-prevention-finasteride-parnes.

19. "Jose Theodore Tests Positive," CBC Sports, last updated February 10, 2006, https://www.cbc.ca/sports/hockey/jose-theodore-tests-positive-1.613274.

Chapter 15

1. Kristen Gurtner et al., "Erectile Dysfunction: A Review of Historical Treatments with a Focus on the Development of the Inflatable Penile Prosthesis," *American Journal of Men's Health* 11, no. 3 (May 2017): 479, doi:10.1177/1557988315596566.

2. Laura Bannister, "The Hard-On on Trial," *Paris Review*, May 18, 2016, https://www.theparisreview.org/blog/2016/05/18/the-hard-on-on-trial.

3. Gurtner et al., "Erectile Dysfunction," 480.

4. Gurtner et al., "Erectile Dysfunction," 479–80.

5. U. Jonas, "The History of Erectile Dysfunction Management," *International Journal of Impotence Research* 13, suppl. 3 (August 2001): S3, doi:10.1038/sj.ijir.3900717.

6. Gurtner et al., "Erectile Dysfunction," 481.

7. Jonas, "The History of Erectile Dysfunction Management," S3.

8. Gurtner et al., "Erectile Dysfunction," 481.

9. Jonas, "The History of Erectile Dysfunction Management," S3.

10. Gurtner et al., "Erectile Dysfunction," 481.

11. Jonas, "The History of Erectile Dysfunction Management," S3.

12. Gurtner et al., "Erectile Dysfunction," 481.

13. Jonas, "The History of Erectile Dysfunction Management," S5.

14. Tim Rausch, "Reclaiming His Throne," *Augusta Chronicle*, April 23, 2007, https://www.augustachronicle.com/article/20070423/NEWS/304239990.

15. Kimberley Hoyland et al., "The Use of Vacuum Erection Devices in Erectile Dysfunction after Radical Prostatectomy," *Reviews in Urology* 15, no. 2 (2013): 68.

16. Jonas, "The History of Erectile Dysfunction Management," S3–S4.

17. Gurtner et al., "Erectile Dysfunction," 482.

18. Jonas, "The History of Erectile Dysfunction Management," S4.

19. Cyriaque Lamar, "An Eyewitness Account of the Most Awkward Urology Lecture Ever," io9, November 16, 2012, https://io9.gizmodo.com/an-eyewitness-account-of-the-most-awkward-urology-lectu-5876545.

20. Kyle MacNeill, "The Story of Viagra, the Little Blue Pill That Changed Sex Forever," VICE, March 29, 2018, https://www.vice.com/en/article/mbxgnx/the-story-of-viagra-the-little-blue-pill-that-changed-sex-forever.

21. John Tozzi and Jared S. Hopkins, "The Little Blue Pill: An Oral History of Viagra," Bloomberg, December 11, 2017, https://www.bloomberg.com/news/features/2017-12-11/the-little-blue-pill-an-oral-history-of-viagra.

22. Michael S. Rosenwald, "The Viagra Jackpot: A History of the Little Blue Pill at 20," *Chicago Tribune*, March 28, 2018, https://www.chicagotribune.com/business/ct-biz-history-of-viagra-20-years-20180328-story.html.

23. Katherine Ellen Foley, "Viagra's Famously Surprising Origin Story Is Actually a Pretty Common Way to Find New Drugs," *Quartz*, September 10, 2017, https://qz.com/1070732/viagras-famously-surprising-origin-story-is-actually-a-pretty-common-way-to-find-new-drugs.

24. Tozzi and Hopkins, "The Little Blue Pill."

25. Rosenwald, "The Viagra Jackpot."

26. Donald J. Nichols et al., "Pharmacokinetics of Sildenafil after Single Oral Doses in Healthy Male Subjects: Absolute Bioavailability, Food Effects and Dose Proportionality," *British Journal of Clinical Pharmacology* 53, suppl. 1 (February 2002): S5–S6, doi:10.1046/j.0306-5251.2001.00027.x.

27. Hossein A. Ghofrani et al., "Sildenafil: From Angina to Erectile Dysfunction to Pulmonary Hypertension and Beyond," *Nature Reviews Drug Discovery* 5, no. 8 (August 2006): 690–91, doi:10.1038/nrd2030.

28. Jacque Wilson, "Viagra: The Little Blue Pill That Could," CNN, March 27, 2013, https://www.cnn.com/2013/03/27/health/viagra-anniversary-timeline/index.html.

29. Megan Garber, "Jagged Little (Blue) Pill," *Atlantic*, March 27, 2018, https://www.theatlantic.com/entertainment/archive/2018/03/20-years-of-viagra/556343.

30. Tozzi and Hopkins, "The Little Blue Pill."

31. Garber, "Jagged Little (Blue) Pill."

32. Garber, "Jagged Little (Blue) Pill."

33. Tozzi and Hopkins, "The Little Blue Pill."

34. Christopher Leone, "Ranking the 10 Most Embarrassing NASCAR Driver Sponsors in History," *Bleacher Report*, May 31, 2013, https://bleacherreport.com/articles/1658364-ranking-the-10-most-embarrassing-nascar-driver-sponsors-in-history.

35. Jim Moore, "Hard Topic, Easy Money; Palmeiro Cashes In on Viagra," *Seattle Post-Intelligencer*, last updated March 12, 2011, https://www.seattlepi.com/news/article/Hard-topic-easy-money-Palmeiro-cashes-in-on-1092712.php.

36. Tozzi and Hopkins, "The Little Blue Pill."

37. Garber, "Jagged Little (Blue) Pill."

38. Garber, "Jagged Little (Blue) Pill."

39. Tozzi and Hopkins, "The Little Blue Pill."

40. James G. Murray et al., "Long-Term Safety and Effectiveness of Sildenafil Citrate in Men with Erectile Dysfunction," *Therapeutics and Clinical Risk Management* 3, no. 6 (December 2007): 977.

41. Wilson, "Viagra."

42. Esther Inglis-Arkell, "Why Viagra Tints Your Vision Blue," io9, February 12, 2015, https://io9.gizmodo.com/why-viagra-tints-your-vision-blue-1685176169.

43. Yun-Yun Li et al., "Visual Impairment with Possible Macular Changes after a High Dose of Sildenafil in a Healthy Young Woman," *International Journal of Ophthalmology* 11, no. 2 (February 18, 2018): 340–42, doi:10.18240/ijo.2018.02.27.

44. Cüneyt Karaarslan, "Ocular Side Effects of Sildenafil That Persist Beyond 24 h—A Case Series," *Frontiers in Neurology* 11 (February 7, 2020): 1–4, doi:10.3389/fneur.2020.00067.

45. John Naish, "The History of Viagra and How It Was Discovered by Accident," *Daily Mail*, last updated December 3, 2017, https://www.dailymail.co.uk/health/article-5134761/A-cure-curse-JOHN-NAISH-wonder-sex-drug.html.

46. N. Mondaini et al., "Sildenafil Does Not Improve Sexual Function in Men without Erectile Dysfunction but Does Reduce the Postorgasmic Refractory Time," *International Journal of Impotence Research* 15 (July 18, 2003): 225–28, https://doi.org/10.1038/sj.ijir.3901005.

47. Garber, "Jagged Little (Blue) Pill."

48. Garber, "Jagged Little (Blue) Pill."

49. Foley, "Viagra's Famously Surprising Origin Story Is Actually a Pretty Common Way to Find New Drugs."

50. "Viagra to Go Generic in 2017 according to Pfizer Agreement," CBS News, December 17, 2013, https://www.cbsnews.com/news/viagra-to-go-generic-in-2017 -according-to-pfizer-agreement.

51. Dawn Connelly, "Three Decades of Viagra," *Pharmaceutical Journal*, May 25, 2017, https://pharmaceutical-journal.com/article/infographics/three-decades-of-viagra.

52. Naish, "The History of Viagra and How It Was Discovered by Accident."

53. "Colombia Dessert," AP Archive, November 11, 2009, http://www.aparchive .com/metadata/youtube/a0af57de6912a45ce5d1c107a6692728.

54. Wilson, "Viagra."

55. Garber, "Jagged Little (Blue) Pill."

56. Foley, "Viagra's Famously Surprising Origin Story Is Actually a Pretty Common Way to Find New Drugs."

57. Nikhil Swaminathan, "Viagra May Give a Boost to the Jet-Lagged," *Scientific American*, May 21, 2007, https://www.scientificamerican.com/article/viagra-may-help -with-jet-lag.

Chapter 16

1. A. Sakula, "Paul Langerhans (1847–1888): A Centenary Tribute," *Journal of the Royal Society of Medicine* 81, no. 7 (July 1988): 414–15, doi:10.1177/014107688808100718, PMCID: 1291675, PMID: 3045317.

2. Diane Wendt, "Two Tons of Pig Parts: Making Insulin in the 1920s," *National Museum of American History*, accessed September 25, 2020, https://americanhistory .si.edu/blog/2013/11/two-tons-of-pig-parts-making-insulin-in-the-1920s.html.

3. V. Malaterre et al., "Oral Osmotically Driven Systems: 30 Years of Development and Clinical Use," *European Journal of Pharmaceutics and Biopharmaceutics* 73, no. 3 (November 2009): 311–23, doi:10.1016/j.ejpb.2009.07.002.

4. R. Daneman and A. Prat, "The Blood–Brain Barrier," *Cold Spring Harbor Perspectives in Biology* 7, no. 1 (2015): 1–24, doi:10.1101/cshperspect.a020412.

5. H. J. van de Haar et al., "Blood–Brain Barrier Leakage in Patients with Early Alzheimer Disease," *Radiology* 282, no. 2 (February 2017): 615, doi:10.1148/radiol .2017164043, PMID: 28099097.

6. M. Dadparvar et al., "HI 6 Human Serum Albumin Nanoparticles—Development and Transport over an In Vitro Blood–Brain Barrier Model," *Toxicology Letters* 206, no. 1 (2011): 60–66, doi:10.1016/j.toxlet.2011.06.027.

7. "OTC Drug Facts Label," U.S. Food and Drug Administration, accessed September 29, 2020, https://www.fda.gov/drugs/drug-information-consumers/otc-drug -facts-label.

8. Jongwha Chang et al., "Prescription to Over-the-Counter Switches in the United States," *Journal of Research in Pharmacy Practice* 5, no. 3 (2016): 149–54, doi:10.4103/2279-042X.185706.

9. "Understanding Unapproved Use of Approved Drugs 'Off Label,'" U.S. Food and Drug Administration, accessed September 29, 2020, https://www.fda.gov/patients/learn-about-expanded-access-and-other-treatment-options/understanding-unapproved-use-approved-drugs-label.

10. Christopher Wittich et al., "Ten Common Questions (and Their Answers) about Off-Label Drug Use," *Mayo Clinic Proceedings* 87, no. 10 (2012): 982–90, doi:10.1016/j.mayocp.2012.04.017.

11. "Developing Products for Rare Diseases and Conditions," U.S. Food and Drug Administration, accessed September 28, 2020, https://www.fda.gov/industry/developing-products-rare-diseases-conditions.

12. Kiran N. Meeking et al., "Orphan Drug Development: An Economically Viable Strategy for Biopharma R&D," *Drug Discovery Today* 17 (August 15, 2012): 660–64, doi:10.1016/j.drudis.2012.02.005, PMID: 22366309.

13. Mary Caffrey, "Claiming 'Orphan' Status, AstraZeneca Sues FDA to Protect Crestor Patent," *American Journal of Managed Care*, June 29, 2016, https://www.ajmc.com/view/claiming-orphan-status-astrazeneca-sues-fda-to-protect-crestor-patent.

14. Brendan Pierson, "Judge Refuses to Block Generic Versions of AstraZeneca's Crestor," Reuters, July 19, 2016, https://www.reuters.com/article/us-astrazeneca-crestor/judge-refuses-to-block-generic-versions-of-astrazenecas-crestor-idUSKCN0ZZ2VW.

15. Ben Elgin, Doni Bloomfield, and Caroline Chen, "When the Patient Is a Gold Mine: The Trouble with Rare-Disease Drugs," *Bloomberg Businessweek*, last modified May 30, 2017, https://www.bloomberg.com/news/features/2017-05-24/when-the-patient-is-a-gold-mine-the-trouble-with-rare-disease-drugs.

16. Walter F. Roche Jr., "Number Deaths Caused by the 2012 Fungal Meningitis Outbreak Underreported," *Tennessean*, December 31, 2018, https://www.tennessean.com/story/news/health/2018/12/31/number-deaths-caused-2012-fungal-meningitis-outbreak-underreported/2447091002.

17. Charles Patrick Davis, "Fungal Meningitis and Steroid Injections: A Health-Care Disease," Medicinenet, accessed September 30, 2020, https://www.medicinenet.com/fungal_meningitis_and_steroid_injections/views.htm.

18. "December 13, 2018: Owner and Four Former Employees of New England Compounding Center Convicted Following Trial," U.S. Food and Drug Administration, accessed September 30, 2020, https://www.fda.gov/inspections-compliance-enforcement-and-criminal-investigations/press-releases/december-13-2018-owner-and-four-former-employees-new-england-compounding-center-convicted-following.

19. "Compounded Unapproved Animal Drugs from Rapid Equine Solutions Linked to Three Horse Deaths," U.S. Food and Drug Administration, accessed October 2, 2020, https://www.fda.gov/animal-veterinary/cvm-updates/compounded-unapproved-animal-drugs-rapid-equine-solutions-linked-three-horse-deaths.

20. "Compounded Unapproved Animal Drugs from Rapid Equine Solutions Linked to Three Horse Deaths."

21. Laurence L. Brunton et al., *Goodman and Gilman's the Pharmacological Basis of Therapeutics*, 11th ed. (New York: McGraw-Hill, 2006), 1794, 1812, 1834.

22. "FDA 101: Dietary Supplements," U.S. Food and Drug Administration, accessed September 30, 2020, https://www.fda.gov/consumers/consumer-updates/fda-101-dietary-supplements.

23. "Questions and Answers on Dietary Supplements," U.S. Food and Drug Administration, accessed September 30, 2020, https://www.fda.gov/food/information-consumers-using-dietary-supplements/questions-and-answers-dietary-supplements.

24. "Is It Really 'FDA Approved?'" U.S. Food and Drug Administration, accessed September 29, 2020, https://www.fda.gov/consumers/consumer-updates/it-really-fda-approved.

25. Martha M. Rumore, "The Hatch-Waxman Act—25 Years Later: Keeping the Pharmaceutical Scales Balanced," *The Pharmacy Times*, August 2009, https://www.pharmacytimes.com/publications/supplement/2009/GenericSupplement0809/Generic-Hatch-Waxman-0809.

26. "What Medicare Part D Drug Plans Cover," Medicare.gov, accessed October 8, 2020, https://www.medicare.gov/drug-coverage-part-d/what-medicare-part-d-drug-plans-cover.

27. "The Medicines Formulary," Guy's and St. Thomas' NHS Foundation Trust, accessed October 9, 2020, https://www.guysandstthomas.nhs.uk/about-us/publications/medicines-formulary.aspx.

28. "Allowed/Disallowed Items," Pharmaceutical Services Negotiating Committee of the National Health Service, accessed October 9, 2020, http://archive.psnc.org.uk/pages/allowed_disallowed_items.html.

29. Jeffrey K. Aronson et al., "Me-Too Pharmaceutical Products: History, Definitions, Examples, and Relevance to Drug Shortages and Essential Medicines Lists," *British Journal of Clinical Pharmacology*, May 13, 2020, 1–9, doi:10.1111/bcp.14327, PMID: 32358800.

30. Aronson et al., "Me-Too Pharmaceutical Products."

31. Aronson et al., "Me-Too Pharmaceutical Products."

32. "World Health Organization Model List of Essential Medicines: 21st List," World Health Organization, 2019, accessed October 8, 2020, https://apps.who.int/iris/bitstream/handle/10665/325771/WHO-MVP-EMP-IAU-2019.06-eng.pdf?sequence=1&isAllowed=y.

33. Rosanne Spector, "Me-Too Drugs: Sometimes They're Just the Same Old, Same Old," *Stanford School of Medicine Magazine*, Summer 2005, http://sm.stanford.edu/archive/stanmed/2005summer/drugs-metoo.html.

34. "Pharmacogenomics Fact Sheet," National Institutes of Health, National Institute of General Medical Sciences, last modified July 13, 2020, https://www.nigms.nih.gov/education/fact-sheets/Pages/pharmacogenomics.aspx.

35. "Pharmacogenomics Fact Sheet."

36. "Drug Safety Communications: FDA Restricts Use of Prescription Codeine Pain and Cough Medicines and Tramadol Pain Medicines in Children; Recommends against Use in Breastfeeding Women," U.S. Food and Drug Administration, last modified April 20, 2017, https://www.fda.gov/media/104268/download.

37. E. J. Stanek et al., "Adoption of Pharmacogenomic Testing by US Physicians: Results of a Nationwide Survey," *Clinical Pharmacology and Therapeutics* 91, no. 3 (March 2012): 450–58, doi:10.1038/clpt.2011.306, PMID: 22278335, S2CID: 21366195.

38. Jocelyn Kaiser, "Is the Drought Over for Pharming?" *Science* 320, no. 5875 (April 25, 2008): 473–75, doi:10.1126/science.320.5875.473, PMID: 18436771, S2CID: 284 07422.

39. Philip Cohen, "Drug-Producing Crops Facing Legal Lockdown," *New Scientist*, March 1, 2003, https://www.newscientist.com/article/dn3436-drug-producing-crops-fac ing-legal-lockdown.

40. Philip Cohen, "GM Crop Mishaps Unite Friends and Foes," *New Scientist*, November 18, 2002, https://www.newscientist.com/article/dn3073-gm-crop-mishaps -unite-friends-and-foes.

41. "ATryn Uses the Power of Recombinant DNA Technology to Offer a Reliable Supply of Antithrombin," rEVO Biologics, accessed October 7, 2020, http://atryn.com /atryn-revo/recombinant.php.

42. Andrew Pollack, "F.D.A. Approves Drug from Gene-Altered Goats," *New York Times*, February 6, 2009, https://www.nytimes.com/2009/02/07/business/07goatdrug .html.

43. H. Luke Shaefer and Analidis Ochoa, "How Blood-Plasma Companies Target the Poorest Americans," *Atlantic*, March 15, 2018, https://www.theatlantic.com/business /archive/2018/03/plasma-donations/555599.

44. C. Lee Ventola, "Direct-to-Consumer Pharmaceutical Advertising: Therapeutic or Toxic?," *Pharmacy and Therapeutics* 36, no. 10 (2011): 669–84.

45. Ann Silversides, "Abramson: Direct-to-Consumer Advertising Will Erode Health Care," *Canadian Medical Association Journal* 178, no. 9 (2008): 1126–27, doi:10.1503 /cmaj.080442.

46. Giampaolo Velo and Ugo Moretti, "Direct-to-Consumer Information in Europe: The Blurred Margin between Promotion and Information," *British Journal of Clinical Pharmacology* 66, no. 5 (2008): 626–28, doi:10.1111/j.1365-2125.2008.03283.x.

47. Zosia Kmietowicz, "New Zealand GPs Call for End to Direct to Consumer Advertising," *British Medical Journal (Clinical Research Edition)* 326, no. 7402 (2003): 1284, doi:10.1136/bmj.326.7402.1284-c.

48. Ventola, "Direct-to-Consumer Pharmaceutical Advertising," 669–84.

49. D. L. Frosch et al., "Creating Demand for Prescription Drugs: A Content Analysis of Television Direct-to-Consumer Advertising," *Annals of Family Medicine* 5, no. 1 (January–February 2007): 6–13, doi:10.1370/afm.611. Erratum in *Annals of Family Medicine* 5, no. 2 (March–April 2007): 179, PMID: 17261859, PMCID: 1783924.

50. Ventola, "Direct-to-Consumer Pharmaceutical Advertising," 669–84.

51. Lee Ventola, "Direct-to-Consumer Pharmaceutical Advertising," 669–84.

52. Dana O. Sarnak et al., "Paying for Prescription Drugs around the World: Why Is the U.S. an Outlier?" Commonwealth Fund, October 2017, https://www .commonwealthfund.org/sites/default/files/documents/___media_files_publications _issue_brief_2017_oct_sarnak_paying_for_rx_ib_v2.pdf.

53. U.S. House of Representatives, Ways and Means Committee Staff, "A Painful Pill to Swallow: U.S. vs. International Prescription Drug Prices," September 2019,

49–50, https://waysandmeans.house.gov/sites/democrats.waysandmeans.house.gov/files/documents/U.S.%20vs.%20International%20Prescription%20Drug%20Prices_0.pdf.

54. Sarnak et al., "Paying for Prescription Drugs around the World."

55. Sarnak et al., "Paying for Prescription Drugs around the World."

56. U.S. House of Representatives, Ways and Means Committee Staff, "A Painful Pill to Swallow," 49–50.

57. Sarnak et al., "Paying for Prescription Drugs around the World."

Epilogue

1. Pablo Uchoa and Yvette Tan, "Covid: What Do We Know about China's Coronavirus Vaccines?" BBC News, July 13, 2021, https://www.bbc.com/news/world-asia-china-57817591.

2. Kathy Katella, "Comparing the COVID-19 Vaccines: How Are They Different?" *Yale Medicine*, September 17, 2021, https://www.yalemedicine.org/news/covid-19-vaccine-comparison.

3. Elie Dolgin, "The Tangled History of mRNA Vaccines," *Nature* 597, no. 7876 (September 16, 2021): 323.

4. Dolgin, "The Tangled History of mRNA Vaccines," 322.

5. Katella, "Comparing the COVID-19 Vaccines: How Are They Different?"

6. Bianca Nogrady, "Mounting Evidence Suggests Sputnik COVID Vaccine Is Safe and Effective," *Nature*, July 6, 2021, https://www.nature.com/articles/d41586-021-01813-2.

7. Philip Ball, "The Lightning-Fast Quest for COVID Vaccines—and What It Means for Other Diseases," *Nature*, December 18, 2020, https://www.nature.com/articles/d41586-020-03626-1.

8. Andrea Shalal and David Lawder, "World Bank Will Boost COVID-19 Vaccine Funding to $20 bln," Reuters, June 30, 2020, https://www.reuters.com/business/healthcare-pharmaceuticals/world-bank-says-will-boost-financing-covid-19-vaccines-20-billion-2021-06-30.

9. Alison Caldwell, "How Were Researchers Able to Develop COVID-19 Vaccines So Quickly?," *University of Chicago News*, February 5, 2021, https://news.uchicago.edu/story/how-were-researchers-able-develop-covid-19-vaccines-so-quickly.

10. Ball, "The Lightning-Fast Quest for COVID Vaccines."

11. Jon Cohen, "Chinese Researchers Reveal Draft Genome of Virus Implicated in Wuhan Pneumonia Outbreak," *Science*, January 11, 2020, https://www.science.org/news/2020/01/chinese-researchers-reveal-draft-genome-virus-implicated-wuhan-pneumonia-outbreak.

12. "Fact Check—COVID-19 Vaccines Are Not a Ploy to Connect People to 5G," Reuters, July 15, 2021, https://www.reuters.com/article/factcheck-covid19vaccines-5g/fact-check-covid-19-vaccines-are-not-a-ploy-to-connect-people-to-5g-idUSL1N2OR2C1.

13. "Selecting Viruses for the Seasonal Influenza Vaccine," Centers for Disease Control and Prevention, last modified August 31, 2021, https://www.cdc.gov/flu/prevent/vaccine-selection.htm#:~:text=The%20influenza%20viruses%20in%20the,circulate%20during%20the%20coming%20season.

14. "Secretary of Defense Austin Issues Guidance for Mandatory Coronavirus Disease 2019 Vaccination of Department of Defense Service Members," U.S. Department of Defense, August 25, 2021, https://www.defense.gov/Newsroom/Releases/Release/Article/2745742/secretary-of-defense-austin-issues-guidance-for-mandatory-coronavirus-disease-2.

15. "Key Parts of Biden's Plan to Confront Delta Variant Surge," Associated Press News, September 9, 2021, https://apnews.com/article/joe-biden-business-health-coronavirus-pandemic-26bace6485d88ad1ae3ef2aea60fbb65.

16. Peter Nicholas, "Why Biden Bet It All on Mandates," *The Atlantic*, September 19, 2021, https://www.theatlantic.com/politics/archive/2021/09/biden-vaccine-mandates/620103.

17. Josh Holder, "Tracking Coronavirus Vaccinations around the World," *New York Times*, September 20, 2021, https://www.nytimes.com/interactive/2021/world/covid-vaccinations-tracker.html.

Index